Hybrid Microelectronic Technology

Electrocomponent Science Monographs
A series edited by D.S. Campbell, Professor of Component Technology,
Electronic and Electrical Engineering Department,
Loughborough University of Technology, Loughborough,
Leicestershire, UK.

Volume 1   *Instabilities in MOS Devices,* by J.R. Davis
Volume 2   *Dielectric Films on Gallium Arsenide,* by W.F. Croydon and E.H.C. Parker
Volume 3   *Ferroelectric Transducers and Sensors,* by J.M. Herbert
Volume 4   *Hybrid Microelectronic Technology,* by P.L. Moran

Additional volumes in preparation

ISSN: 0275–7230

This book is part of a series. The publisher will accept continuation orders which may be cancelled at any time and which provide for automatic billing and shipping of each title in the series upon publication. Please write for details.

Hybrid Microelectronic Technology

Peter Moran
*Senior Scientist at the
National Microelectronics Research Centre,
University College
Cork, Ireland*

GORDON AND BREACH SCIENCE PUBLISHERS
New York   London   Paris   Montreux   Tokyo

Copyright © 1984 by Gordon and Breach, Science Publishers, Inc.

Gordon and Breach Science Publishers
One Park Avenue
New York, NY 10016
United States of America

42 William IV Street
London, WC2N 4DE
England

58, rue Lhomond
75005 Paris
France

P.O. Box 161
1820 Montreux-2
Switzerland

48-2 Minamidama, Oami Shirasato-machi
Sambu-gun
Chiba-ken 299-32
Japan

Library of Congress Cataloging in Publication Data

Moran, Peter, 1948–
 Hybrid microelectronic technology.

 (Electrocomponent science monographs, ISSN 0275-7230; v. 4)

 1. Hybrid integrated circuits. I. Title. II. Series. TK874.M53355 1984 621.381'73 83-25484 ISBN 0-677-06560-4

Library of Congress Catalog Card Number 83-25484. ISBN 0-677-06560-4, ISSN 0275-7230. All rights reserved. No part of this book may be reproduced or utilized in any form or by any means, electronic or mechanical, including photocopying, recording, or by any information storage or retrieval system, without permission in writing from the publishers. Printed in The United Kingdom by Bell and Bain Ltd., Thornliebank, Glasgow.

Introduction to the Series

This series of monographs in the field of electronic and electrical component science and technology is composed of authoritative reviews written by acknowledged experts. Reviews of this type are normally collected in volumes containing several chapters on different, often only distantly related subjects; it is not usual for subscribers to want all the reports which are bound into one volume. This situation limits the availability of each review due to the cost of the complete volume. The present series is aimed at solving this problem by publishing each review as a separate monograph.

In addition to volumes concerned with particular components, this monograph series will include reviews on associated material, particularly with regard to other interconnection technologies, quality control, failure analysis, and product safety.

The publications are designed for use by final-year undergraduates, postgraduates, and engineers working in the field. It is hoped that the books will prove stimulating and useful.

D.S. Campbell
Loughborough University of Technology

Contents

		Page
	Preface	1
1	INTRODUCTION P. L. Moran	4
2	BASIC THICK FILM PROCESSING J. C. Taylor	19
3	THICK FILM PASTES AND SUBSTRATES B. Walton	41
4	ADD ON COMPONENTS AND ATTACHMENT METHODS J. R. Polden	53
5	THIN FILM PROCESSING H. T. Law	64
6	THE DESIGN OF THICK FILM HYBRID CIRCUITS I. D. Salisbury	82
7	PACKAGING HYBRID CIRCUITS B. C. Waterfield	106

		Page
8	AUTOMATING THE PRODUCTION OF HYBRID CIRCUITS G. W. Griffiths	124
9	THE APPLICATION OF HYBRID TECHNIQUES N. G. Burrow	136
10	A CUSTOMER'S VIEW OF HYBRID TECHNOLOGY F. N. Sinnadurai and A. Saunders	157
11	QUALITY CONTROL AND ASSURANCE IN HYBRID CIRCUIT PRODUCTION J. T. Lynch	183
	References and Bibliography	206
	About the Authors	215
	Index	220

Preface

The United Kingdom Chapter of the International Society for Hybrid Microelectronics has been arranging Professional Advancement Courses in Hybrid Microelectronic Technology for several years. The aim of these courses is to disseminate information about the techniques and industry to as wide an audience as possible, but in particular to potential users of hybrid microelectronics as well as those already or newly involved. The lecturers on the course are all recognised experts in their own specialization and are practicing engineers, researchers or managers in the industry.

Over the years there has been a steady demand for copies of the notes that accompanied the course from people who were unable to attend (all the courses have been over subscribed) and in order to satisfy this demand the lecturers have expanded their notes into a form suitable for publication. This book should therefore be of interest to engineers, both old and new, in the hybrid industry, users and potential users of the technique who wish to gain a deeper insight of the industry and to engineering students who would like to broaden their knowledge of this rapidly growing area of microelectronics.

The general theme of the course has not been to present the mechanics of production but more to emphasise the reasons for using the technology and the general background to the industry. Hence there is little information in this book about the details of how to print, fire and trim thick film circuits since these have already been covered adequately in other texts (some of which are listed in the bibliography), and can be seen in operation at exhibitions or on a visit to a manufacturer.

In order to make reasoned decision on whether to use the technology in a particular application, information on electrical performance, reliability, quality control, other successful application areas, security of supply and manufacturing difficulty is important and the main objective of this book is to supply, as far as is possible in text form, some of this background. Therefore as well as chapters on the basic thick film technique, thick film materials and components, which are necessary to describe the process, there are chapters discussing the design of thick film circuits and their packaging (which forms one of the major cost-performance decisions in the process). Both the design parameters and also the package sealing techniques used are covered. A chapter has been included on thin film processing since this finds application in a number of specialist areas and, apart from the basic deposition and pattern delineation technique, has similarities with thick film methods.

One way of lowering costs is through the use of automation. However automation usually implies high volume production; and yet the thick film process is generally important in low to medium volume applications. One chapter therefore discusses the philosophy and practice of automation in hybrid technology. The film hybrid is an assembly of components and the integrity with which the devices are assembled is of critical importance. Hence a chapter is included on the vital subject of quality control and reliability assessment. Another chapter is concerned with applications of hybrid technology both in the general sense of how to assess whether a certain application is likely to be successful and also by describing some specific reasonably high volume applications that take advantage of particular aspects of the techniques. Finally the technology is looked at from the viewpoint of one large user of hybrid methods and the reasons why they have adopted it in a number of areas are examined. Also covered are the results of some of their studies into long term reliability and the ways and reasons why they have tried to influence the industry to move in certain directions.

Each chapter in this book has been written by a different

Preface

contributer who has been asked to interpret the technology as it affects his own area of expertise. Hence a certain amount of duplication occurs in the text as some devices or materials are discussed from a number of different viewpoints. Rather than attempt to edit out this repetition it has been left in as an indication that the particular device or material confers several different advantages on the technology.

Attempting to present a study of a high technology industry is similar to photographing a moving object; by the time the picture is developed, the scene has changed. However there seems to be sufficient continuing new interest in using hybrid microelectronics and as the industry is expanding, sufficient new people working with and using the technology to make such a task worthwhile.

1
Introduction

P. L. MORAN

The technology of microelectronic engineering has made very considerable advances over the past decade and a half and the acceptance of the technology is such that it is now difficult to imagine a world without electronics. The extent to which electronics has become part of everyday life can be gauged from the recent estimates that the market for electronics equipment in the U.S.A., Japan and Western Europe will exceed \$M200,000 in 1982, a sum that equates to several hundred dollars for every individual. These products vary from at one extreme low cost high volume consumer products, such as video games, to highly specialised and costly aerospace circuits at the other and in between there is an apparently unlimited variety of applications all with their own requirements of quantity, cost and reliability. The one common component in these products is of course the silicon integrated circuit and with only relatively minor differences, the same integrated circuit could be used in all or any of these application areas. It is interesting therefore to ask why there should be a such wide variety of cost, volume and performance parameters relating to a component that appears in near identical form in each application.

In order to examine, or at least begin to examine, this apparent contradiction it is first of all necessary to study the silicon integrated circuit and its basic features and subsequently to look at how it might be used in a number of applications. With this background it should then become clearer why thick and thin film technologies have such an important supporting role to play.

Introduction

There are always several different ways of meeting any particular objective; if the demand for a certain application is relatively low, then it probably does not matter which is chosen provided the essential engineering features are satisfied. However, when the demand is high and relative costs, performance, reliabilities, security of supply, etc. become important, one method will begin to dominate and this is now the situation in which thick film technology finds itself since its long term acceptance in large volume telecommunications, computer and automotive systems appears assured. Its relative performance compared to competitive techniques is also therefore of relevance as is the history of the technology that maps the series of market changes that have motivated the developments of the technology and have put it into such an important position.

The technology of silicon integration has made many advances in the last decade. The roles of bipolar TTL and ECL, whilst still significant, have become less dominating and the use of CMOS and NMOS circuitry has increased remarkably as new techniques of processing have been introduced. The improvements in device resolution and processing techniques have led to increases in device complexity and yields. However, to the user, apart from this greater complexity in circuit function and also lower prices, there have been remarkably few changes and those that there have been are of scale rather than technique. The silicon integrated circuit is made by the successive addition of carefully controlled layers in equally carefully controlled positions. The typical device dimensions in current use vary from about 3µm to about 8µm in the plane of the silicon surface and are measured in fractions of a micrometre normal to the surface. The purity of the materials used is often specified at better than one part in 10^7. These dimensions and purities are such that even the smallest amounts of contamination can have drastic effects on the performance and yield and a prime requirement of any silicon manufacturing plant is to maintain the highest possible standards of cleanliness and purity. The equipment used in silicon processing is complex and requires very close tolerance machining to meet the dimensional accuracy necessary in modern devices. It is not surprising therefore that such equipment is extremely costly. Many of the materials used in the processing are, apart from the basic silicon, toxic, inflammable or explosive or are powerful acids and therefore an adequate standard of safety engineering is essential.

It can be seen that prior to any marketing considerations, a very large capital investment is necessary to procure the equipment and building and a reasonably large overhead is required simply to keep the production line running. It is imperative therefore that the efficiency with which the equipment is utilized should be as high as possible and very large scale volume production is a pre-requisite for profitable operation.

High volume by itself is not a guarantee of success. The typical yield of a large integrated circuit may be as low as 20%. Yield is affected by many parameters, the quality of the photomasks, the purity of the chemicals and equipment, the quality of the silicon, the contamination in the atmosphere, and not least the skill of the design engineer. However, the probability of incorrect functioning appears to be a near random event and a device will fail to function if there is just one fault point contained within it. The pressure therefore is to reduce the device area to a minimum both to maximise yield and also to increase the number of devices manufactured per slice. Smaller devices also give rise to higher operating frequencies. A typical slice is shown in figure 1.1, containing large scale integrated circuits. This reduction in area will be beneficial until the device features become so small that they cannot be manufactured with the available equipment. If a 10% reduction in area improves the yield by around 10% (i.e. say from 20% to about 22%) and produces a 10% increase in the number of devices manufactured, then there is clearly considerable advantages to the producer in reducing the device area to an absolute minimum.

Large integrated circuits are very complex and the amount of engineering effort needed to design them is commensurately high. A typical large scale custom integrated circuit might require three to four man years of work. A device that is destined to become a standard company product and, therefore very price sensitive, considerably more, and a very high volume large scale device such as a memory circuit much more effort still. The design cycle could be in excess of two years (certainly over one

Figure 1.1 A silicon slice containing LSI devices.

year) and in order to employ a fully custom engineered integrated circuit a degree of assurance that the product will have a good market over which the development costs can be amortized is required. And yet as the product will not be available for at least two years the risk that either the customer's requirement will have changed or that a competitor will produce a similar device in the meantime is also significant.

A typical large scale integrated circuit might contain 10,000 or more transistors and perhaps forty to sixty input and output connections. It is obvious that it is not possible to repair faulty devices by transistor replacement, although some designs do allow for built-in redundant sections to be used instead of faulty sections but the incidence of this is not very great. With so few connection points to such a large amount of circuitry there is clearly a very difficult testing problem. It is impossible to carry out a 100% functional check on such a device and the usual method of testing is to examine only those functions and input sequences of direct interest. Even this limited testing is expensive and produces much more added value than the basic device is worth. However, it will be recalled that the device yield might only be in the region of about 20% and thus the task is to seek out the good device rather than the classic testing and quality control problem of rejecting the bad component. The circuit will generally undergo two testing operations. The first will be carried out while the device is still part of the overall slice and here a simple check of certain functional parameters will be made and obvious reject devices marked. It is important not to package faulty devices since the cost of packaging is high. The second test, made after packaging, is to determine the precise parameters of the circuit. Quite often several versions of a circuit will be marketed each with certain performance; this performance being determined at the final testing stage. One important consequence of this for hybrid techniques is that it is difficult, if not impossible, to use parametrically tested unpackaged integrated circuits.

Silicon integrated circuits have remarkably few failure modes provided they are used with the correct electrical parameters, kept free from contaminants and within their design operating temperature. The main functions of the integrated circuit package are to protect the device from hostile environments, to improve the dissipation, to provide a more usable connection to the device (the pad areas are kept to a minimum to maximise the utilization of the silicon) and to make the device more amenable to automatic handling. The device identification is also stamped on the package. A selection of device packages is shown in Figure 1.2 together with some unpackaged devices and one of the more obvious problems of packaging immediately becomes apparent. The area utilization is very poor and becomes poorer as the number of input and output pins increases. The package itself is also expensive and the connection of the integrated circuit to the package quite difficult. Packages with more than about forty to fifty pins are exceedingly cumbersome and are not really amenable to subsequent automatic assembly.

The performance of silicon integrated circuitry depends on the type of circuit under consideration. Digital CMOS circuits give rise to very low power and reasonably high speed operation, although the power dissipation is related to the actual clock frequency used. Digital ECL circuits can operate at about 0.3ns propagation delay, but do consume large amounts of power-sufficient to give them a dissipation problem. Analogue performance on the other hand is rather more limited since the production variables encountered are fairly large. Transistor gains can only be predicted with a tolerance of +100%, -50%, although device matching or ratioing is usually better than ±2%. Even this value is inadequate for a number of modern linear circuit applications and the usual method of operation is to use the integrated circuit to provide the active gain and passive feedback to define the overall performance. Any energy storage requirements would generally be provided by external passive components.

In summary therefore the silicon integrated circuit provides excellent device performance at a remarkably low cost provided the production volumes are large enough to amortise the design costs and to allow the establishment of fully automated handling, testing and packaging equipment. The cost per transistor connection is extremely low and becomes lower as more transistors are included on the circuit provided the yield remains acceptable. However this increased complexity brings a host of new problems including extra design effort, more difficult testing both pre and post packaging, and longer design cycles. Historically it has

Figure 1.2 A selection of integrated circuit packages together with the chips contained within them.

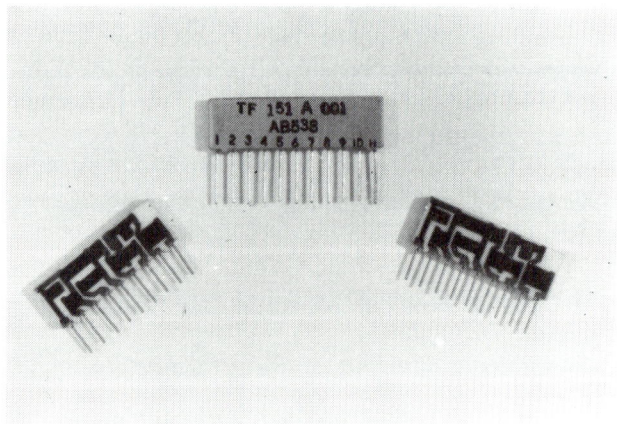

Figure 1.3 Some typical thick-film resistor networks (courtesy A. B. Electronic Products)

Introduction

always been predicted that the number of input and output connections to an integrated circuit would initially rise as complexity increased and then fall as the entire function became available within one or two devices. This trend has certainly been noticed in simple programmable systems where the function of a device can be altered by either a mask programming or by external programming. However, human nature being what it is, once the cost of systems began to fall through the use of large scale integrated circuits, more complex requirements were placed on the system, thus increasing the amount of circuitry required. Similarly it became more cost or performance competitive to use circuitry rather than programs to carry out certain functions. As the systems have become more specialised, the volume requirements for many of the circuits has dropped to the point where the design costs and times of a custom designed integrated circuit are not acceptable. In order to overcome this problem three distinct trends are discernible. Firstly semiconductor manufacturers are offering a fabrication service of customer supplied designs thus relieving the manufacturer of providing the design overhead. Secondly the use of gate arrays is increasing. In a gate array the process is standard up until the final stages of interconnecting the individual gates (or components in the case of an analogue circuit). In this technique only one (or possibly three, for multi-layer metallization) masking layer is particular to any one design. Design costs and times are therefore very much reduced but at the expense of poorer utilization of the silicon area, since not all the gates will generally be used. Thirdly, as a compromise between gate arrays and fully custom devices, the cell array has been developed. Here a library of tested circuit functions or cells is available and any particular circuit design is arranged to be an interconnection of standard cells. Thus better utilization of the silicon is obtained compared to the gate array but at the expense of requiring a larger number of masks and also requiring unique fabrication (i.e. unlike the gate array in which only the final stage is unique).

Neither of these techniques are as efficient in their use of silicon as the fully custom device but they are amenable to much more automation in the design process. Computer programs are now available for the fully automatic layout of both gate arrays and cell arrays (i.e. the use of interactive graphics is no longer essential, although they will still have a useful role to play). The result of these trends is that the design costs and times of "semi custom" integrated circuits are reduced significantly and that alternative sources of the supply of custom devices are becoming available. However another consequence of the increased use of higher gate count random logic semicustom devices is a need for an increased number of input and output connections to the device. The current figures mentioned are for several hundred connections for 10,000 gate circuits. There are thus many considerations other than the electrical function to be taken into account in the construction of electronic equipment and one of the more important is the choice of interconnection technology. It is in this area that significant differences in equipment practice are found, and it is against this current and historical background of silicon technology that the use of thick film technology needs to be assessed and compared to printed circuitry and thin film technology as an interconnection method. The historical developments are important since they explain why particular investments were made and only certain products are currently popular even though there might seem to be better ways now of producing them. It is not easy for a manufacturer to scrap a large investment and they will make every attempt to continue employing developed processes for as long as they remain profitable. As experience develops, confidence in the technology grows and its performance over a real (as opposed to accelerated) life becomes known. Technology therefore assumes a momentum within which changes tend to be relatively gradual.

Printed circuitry traditionally required the use of components with wires or pins that were placed through holes in the board and a soldered connection made to the copper tracks. The conductors are formed by a subtractive photolithographic etching process from the starting material of copper clad fibre glass or similar material. The tracks could be either on just one side or both sides with connections between the sides made via metallized holes. The through holes are invariably drilled individually and subsequently plated. The use of several layers is also possible with the various layers being laminated together to form the final wiring assembly. All electrical components are added as discrete entities and a degree of compliance in the leads is necessary to allow for the difference in thermal coefficient of

Introduction 13

expansion of the board and its components. Automatic insertion is possible provided the number of leads is not too great (the precise number depends on the user and equipment but the upper limit does appear to be about forty). The resolution of the printed wiring board is governed mainly by the size of the drilled through connection and the surrounding copper area, not by the track dimensions. For all these reasons the use of printed circuitry generally implies a large assembly and to minimise costs the incidence of connectors must be kept as low as possible. Therefore the trend is to put as much circuitry onto one board as possible and the result is a large printed wiring board with a similarly large number of individual components and there is no intermediate step for testing between the devices and the fully populated board. The pressure to reduce cabinet and metalwork sizes and hence costs must also be recognised and the extent to which printed wiring boards can do this by themselves is limited.

Thin film technology is based around the evaporation or sputtering of either resistive or conductive materials onto a polished glass or ceramic surface. The technology is difficult and obtaining consistent results not easy. The patterns are formed by subtractive etching in much the same way that printed circuit boards are produced but much higher resolution can be obtained. The usual method of manufacture is to purchase plates already coated with a resistive layer and conductive layer from a specialist supplier and to remove selectively the conductive and resistive layers. The various resistor values are formed by changing the geometry. Unlike the printed wiring board, components (i.e. resistors) are formed as part of the process and with careful processing and laser trimming, very high precision, high stability resistors can be obtained. Active devices, either transistors or integrated circuits, and capacitors are added later and are mounted on the surface of the substrate. The semiconductor components historically were unpackaged and the connections made directly from the integrated circuit to the thin film conductor. Two very important advantages were therefore obtained; firstly an increase in reliability due to the reduced number of multimetal connections and secondly a reduced size for the given circuit function. The disadvantage is that it is now necessary to package not just one device but the entire assembly in a high integrity package to prevent contamination. The thin film materials themselves are susceptible to degradation by exposure to ionic contaminants and mechanical damage.

The thin film process is generally presented as a batch flow method and therefore rather difficult to automate. This view needs to be treated with some caution however as it is probable that there has never really been the market pressure to automate since thin film products have typically been designed into aerospace, medical and similar products where the cost of the basic substrate and components is not particularly significant and the production volumes are relatively low.

Thick film technology, which will be described in much greater detail in subsequent chapters, has either a very long history or a shorter history than any of the other interconnection techniques depending on where one wishes to define the beginning. The idea of selective deposition by printing is obviously very old, as is the use of materials fired onto ceramic. However, it was probably not until the early to mid-1960's that the idea of making a circuit in such a manner using a ceramic substrate began to take on real significance and as in so many developments it was IBM who motivated the general acceptance of thick film processing by designing it into their 360 range of computers. (Although Centralab had been using the technique for some while). From the rest of industry's point of view it was their use of resistor networks that was of most significance. Some typical IBM networks are shown in Figure 1.3. Their consumption was, and still is, enormous running into hundreds of millions of units per annum and the development of automatic handling methods and improvements in the materials became essential. With so many devices in field use, real reliability data was available (if not to the general public, then at least the continued acceptance implied no problems). Through their purchases from subcontractors the technology spread to other industries and it is interesting to speculate that had IBM chosen to use thin film resistor networks, whether the thick film process would have arrived where it is today.

With such an important large scale customer (at that time IBM controlled probably 75% of the world's computer market) adopting the technology other developments were bound to follow. It was soon found that good quality gold conductors could be manufactured and unlike the thin film process there was no wastage through etching. The resistor materials gradually improved and whilst they are still not as good as the best thin film components, they are more than adequate for most applications. Laser trimming, which is a very high speed operation, both for resistor value and circuit function tended to eliminate the need for adjustable resistors. The gradual availability of insulating thick film materials meant that first of all simple crossovers and subsequently full multi-layer assemblies were manufacturable - an important advantage over thin film techniques. The range and performance of thick film resistors gradually improved until it is now difficult to find an application that cannot be met by them and the processing gradually became easier. Typical temperature coefficients are now ± 50 ppm/ $^\circ$C and 0.1% per decade hour drift. Resistors are routinely trimmed to ± 1% tolerance.

Historically therefore the major motivation for the thick film process was resistor networks for use in transmission line termination and load resistors combined with very small scale integrated circuits. The ruggedness and (relative) ease of automation of the technique allowed for high volume production. The networks were assembled together with the packaged active devices on a printed wiring board. Gradually the space and reliability advantages of using unencapsulated integrated circuits, particularly in aerospace and medical applications, either as multidevice assemblies or combined with resistors and capacitors became to be appreciated. A typical high reliability circuit is shown in figure 1.4. This complete functional entity could then be tested prior to assembly on a printed wiring board. The acceptance of thick film techniques into consumer and industrial applications was rather slower arriving mainly because of the expense of handling the unencapsulated devices and the fact that semiconductor manufacturers often charged more for the unpackaged device than the packaged one. In many early thick film industrial circuits, conventional wire ended components were attached through holes in the substrate - simply because that was the way components were attached to printed wiring boards and it was known to be reliable. Gradually however the advantages for automation of placing the components on the surface became to be appreciated and more or less at the same time active components became

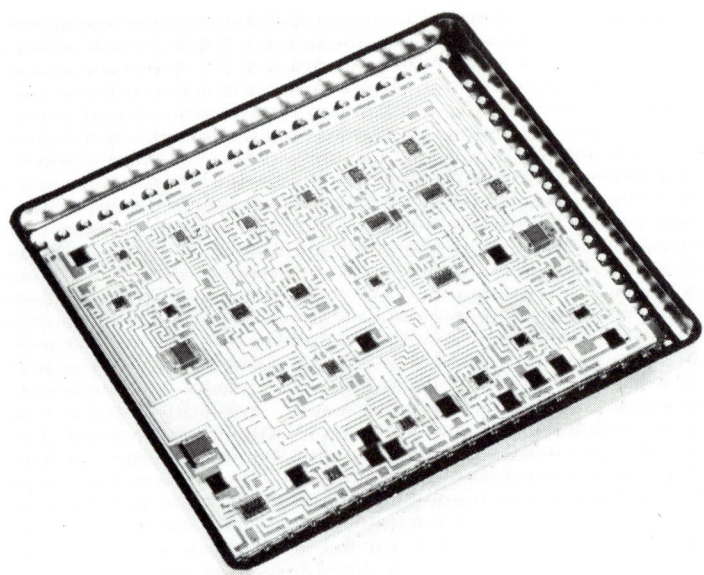

Figure 1.4 A typical high reliability multilayer hybrid circuit (courtesy The Plessey Co.)

Figure 1.5 A typical surface mounted hybrid (courtesy The Plessey Co.)

Introduction

available that were specifically packaged for this so called surface mounting method. Looking back it may seem an obvious and simple step, but it was a very bold market decision to tool up to supply a large range of components in a totally new package style and much credit must go to the Philips organisation for their pioneering work. A typical industrial grade surface mounted hybrid is shown in Figure 1.5.

More or less simultaneously with the emergence of the low cost surface mounting devices, three other market forces were becoming apparent. The relative cost of the electronic devices was falling rapidly compared to the costs associated with the cabinets, connectors, etc. and there was a need to reduce the size of the electronic assemblies. Secondly, the complexity of the electronic assemblies was becoming such that it was necessary to break down the circuit functions into less complex subassemblies and yet still retain the use of a single (or at least very few) printed circuit boards. Finally it was becoming obvious that conventional semiconductor packaging techniques were just not adequate for high pin count integrated circuits and it was found that pinless surface mounting packages were more suitable for the higher number of connections. As these packages were made from ceramic and no compliance between the package and substrate is possible, it seems that for the larger devices at least, only a ceramic based interconnection technology is usable. There are however developments in hand to produce surface mounted devices with such compliance and printed wiring boards whose thermal expansion matches ceramic.

For all these reasons it can be seen that the thick film process has a role to play in almost every aspect of the interconnection of silicon devices. The products range from simple but high performance and compact resistor networks, through reasonably complex, yet complete functional entities, industrial grade products, to very complex, miniaturised high reliability circuits. Although the disciplines involved in each of these product areas are very different, many of the materials and processes are the same. In the past thick film technology has been used exclusively for networks and to obtain the ultimate in miniaturisation by interconnecting a number of unencapsulated small scale integrated circuits. Some specific applications, for example analogue to digital converters and heart pacemakers, used the ability to combine active devices with high performance resistors (and, if required, good quality capacitors) to obtain a complete functional entity inside one package. The arrival of miniature packaged active devices enabled hybrid assemblies to be price competitive with printed wiring techniques and it became economically feasible to use the printed wiring board to interconnect a number of reasonably complex sub-assemblies each of which is a complete functional entity. The consequence of this concept is that the printed wiring board becomes simpler, and usually, smaller for a given circuit function.

The future of interconnection techniques is currently the subject of much discussion. For low to medium pin count integrated circuits, current methods are adequate. However, within a short time high pin count gate and cell array integrated circuits are going to be a fact of life and the currently used methods are simply impractical for such devices. There is therefore much room for innovation and the flexibility of the thick film process would appear to be invaluable in the evolution of new interconnection ideas.

2
Basic Thick Film Processing

J. C. TAYLOR

INTRODUCTION

Film technology is the generic name given to an electrical interconnection technique based on "thick film" or "thin film" methods. When either of these are combined with active add-on components the result is called a "hybrid" circuit. The labels "thick" and "thin" detail the method by which the conductive, resistive or insulating films are deposited onto the substrate (glass, alumina, quartz etc.) and also the film thickness. Thin film involves the use of an evaporation or sputtering process performed in vacuum, whilst thick film technology uses screen printing and subsequent processing of special "paste" materials.

HISTORY

The origins of thick film date back some 3,000 years (compare this with solid state electronics which has had a life so far of only 30 years) and was developed from the method of "silk screen" printing used by the Chinese. The very fine silk threads comprising the mesh were ideal for depositing multi-layered coloured patterns onto fabrics and it is interesting to note that even today a large amount of screen printing is carried out in the field of graphic art and decoration. Metallizing of ceramic materials probably began somewhere in the 19th Century; however the earliest work which has greatest relevance to thick film today was done by Pulfrich, in the 1930's (1,2). In this work, vacuum tight seals were fabricated by coating ceramic with a finely distributed iron and molybdenum powder and then firing the samples at high temperature in a controlled atmosphere. Pulfrich's conclusion was that materials control was fundamental to successful results and this is more true today than it ever was. The thick film technique as it is currently used for electronic circuitry developed from work in the mid-1950's when the transistor was beginning to be applied more widely. After the second world war, conductor pastes consisting of silver powder, glass powder and petroleum jelly were being fired onto ceramic materials. Today thick film is a highly developed mature interconnection technique which allows integration of passive components as part of the process - resistors, capacitors and inductors can be incorporated as the application demands.

Thin films on the other hand do not have such an extensive history. The first thin films, reported to have been produced by Faraday in the 1850's, were made by exploding metal wires in an inert atmosphere. Evaporated films have found many industrial uses as the quality of vacuum equipment has improved. These applications range from anti-reflective coatings and mirrors to electronic circuits and components and thin film techniques are of course the method by which aluminium is deposited onto semiconductors. Thin films tend to be used where the ultimate in line definition is required and hence its application at microwave frequencies. True multilayer thin film circuits have not found widespread use due to the need for evaporation or sputtering of insulating materials and their subsequent subtractive processing. An example of an early thin film circuit is shown in figure 2.1.

Thick film multilayer on the other hand is a well established and widely used technique for both simple and complex hybrid circuits, and an example of space saving that can be achieved is demonstrated in figure 2.2, where a system is realized using both PCB and hybrid technologies.

With the advent of semiconductors, film technology was seen as a replacement for the more conventional printed wiring board approach. However in practice it has turned out to be more of a complementary interconnection technique, with application dictating the use of one approach or the other, or often both. As with all new technical innovations, the development is quite

Basic Thick Film Processing 21

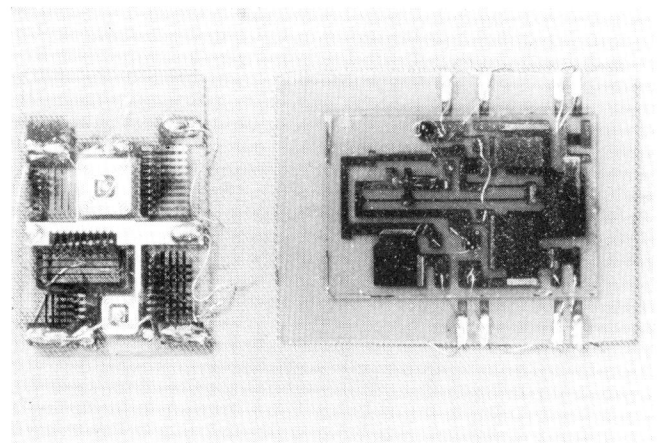

Figure 2.1   An early thin film circuit.

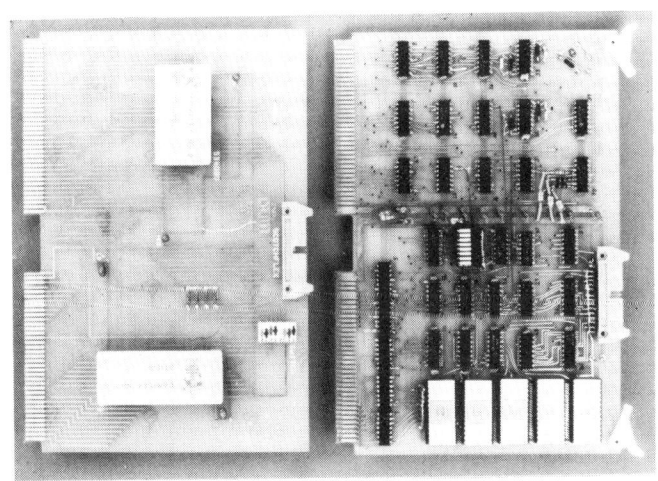

Figure 2.2   An example of the space saving that can be achieved using thick film technology. Both circuit boards perform the same function. Note also the much simpler printed circuit board.

often funded by those who tend not to have a broad interest in commercial and economic factors - e.g. military and aerospace. However in the case of film technology economic factors favoured the success of the technique.

To confirm this it is interesting to note the market developments of film circuit technology in the U.K. Figures from 1979 to 1980 estimated a 33% growth rate (excluding inflation) in total market value. This is significantly better than those for printed wiring boards (estimated 11.5%) which is a more mature market, with perhaps some over-capacity (4) and the PWB market has been eroded to some extent by this significant growth rate in film technology usage over the last 3 to 4 years.

The remainder of this chapter will be devoted to thick film technology, which at present sees far greater usage compared to thin film. The reader interested in thin film will find reference (3) and chapter 5 useful.

WHY HYBRIDS ?

Before the basics of processing can be considered it is necessary to ask the question WHY HYBRIDS ? Clearly the answer to this question will depend on the specific application, type of circuit, life etc. In general the benefits of using hybrids to form system or sub-system elements can be grouped under the following headings.

Miniaturisation

As a key benefit the size and weight of an electronic module can be reduced due to the high packing density available (e.g. man-portable systems) or, as in some cases (e.g. avionics) more electronics can be packed into an equivalent volume.

Performance

By minimising the length of interconnect between devices propagation delays can be reduced thus improving circuit operating speed particularly with high speed logic (ECL). Stray capacitance and inductance can also be reduced which may be of use in rf circuit design and construction.

Versatility

Hybrid microelectronics offers the system engineer an extremely versatile interconnection technology by virtue of the mix of semiconductor devices that can be incorporated (e.g. linear, bipolar logic, MOS, CMOS etc.) together with a wide range of passive lumped and distributed components which would be extremely difficult to fabricate in monolithic form as a single entity.

Integration

Many different circuit functions can be integrated on a single substrate by utilising the versatility already mentioned. As far as the user is concerned the hybrid can be treated as a monolithic device in terms of its connection to the outside world. The hybrid will never yield to the monolithic device as some suggest since it will always be possible to integrate a number of complex chips into a hybrid together with those components that cannot be integrated such as high stability resistors.

Reliability

The hybrid is reliable by virtue of the minimisation of inter-connections (wire bonds, soldered joints etc.) particularly where the chip and wire approach (naked semiconductor IC's mounted on a thick film substrate and electrically connected by wire bonds) is used. There is, in effect, a lower component count at the board (PCB) level which results in reduced board complexity. Sensitive components may be hermetically sealed within the hybrid thus providing protection against hazardous environments.

BS Approval

A manufacturer who has a BS9450 approval is able to supply hybrids (not necessarily all types) to an assessed quality. This approval states that his production facility (i.e. his capability) is qualified to the defined standard. This standard is particularly relevant to hybrids procured for professional applications.

PROCESSING

Before any circuit can be produced the electronic design based on the circuit diagram has to be translated into a physical layout. Once the decision has been taken as to which construction method to adopt (single layer, simple cross-over, multilayer, chip-and-wire, soldered pre-packaged components etc.), a set of design rules is used to prepare the layout, and dictate the overall substrate size (5).

The object of the design layout exercise is to produce a set of artworks from which the substrate can be fabricated. This operation is analogous to laying out a printed circuit board (an etched copper/glass fibre laminate used to interconnect discrete components). In addition however to the routing of interconnects between devices, dielectric areas must be defined for crossovers (to provide electrical isolation) and resistor dimensions must be detailed taking into account value, tolerance, stability, power dissipation and terminating conductor material.

Drawings of the layout are produced at this stage for all subsequent fabrication, assembly and testing operations. CAD (Computer Aided Design) can be used to speed up the design process and produce all manufacturing documentation.

A cut and peel Rubylith artwork is produced by hand or machine for each layer of the circuit, scaled typically to five times final size in order to obtain sufficient accuracy for registration between each layer. Rubylith is a two-layer plastic (red and clear typically) laminate and areas of the red plastic are removed to define the pattern. Photographic reduction is used to produce the master artworks on high contrast film.

An alternative method for producing artworks directly, the photoplotter, is now seeing more widespread use. The photoplotter can accept dimension information from the CAD system and translate this information into the movement of an X-Y table. Patterns defined on a mask are projected as appropriate onto high contrast film mounted on the X-Y table. Hence the master artworks can be produced automatically by "writing" the necessary shapes directly onto film. A typical set of photomasks

Basic Thick Film Processing 25

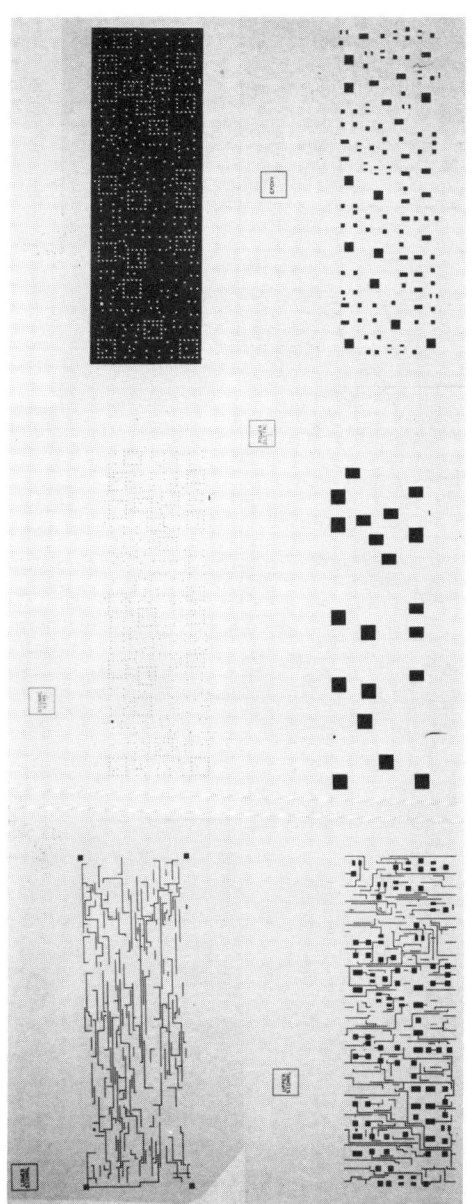

Figure 2.3 A typical set of photomasks for a complex multilayer hybrid.

is shown in figure 2.3.

Thick film substrates are fabricated by a screen printing process but before printing can be undertaken a screen defining each of the patterns must be produced. Screens are prepared in the following manner. A woven stainless steel or polyester mesh is stretched over a rigid screen frame; the tension of the mesh being controlled. A two-part epoxy adhesive is used to adhere the mesh to the frame. The filament diameter and spacing of the mesh is determined by the type of paste to be printed. For example 325 mesh stainless steel cloth (325 wires per inch) would be an appropriate choice of mesh for a gold conductor print where good edge definition is required. In this case the wire diameter is 30μm with a spacing between wires of 50μm (typical). Resistor and dielectric materials are printed typically using 200 and 280 mesh cloths respectively where in general thicker prints are required.

The pattern is defined on the screen using a photographic process. A common technique known as the direct/indirect method makes use of an ultra-violet light sensitive diazo type dry film emulsion which is adhered to the screen mesh using a liquid sensitizer. Following drying the photo-positive master artwork produced earlier is placed emulsion side down (for good definition) onto the 21μm (typically) thick emulsion which is then exposed through the artwork to UV light. The unexposed parts of the screen are washed out with warm water. After inspection and drying the screen is ready for use. Figure 2.4 shows the emulsion coating process and figure 2.5 a completed screen.

Alumina (96%) is the most common substrate material chosen for its good dimensional stability, its low thermal impedance (compared to pcb glass fibre laminate) and its property of being chemically inert particularly at the high firing temperatures experienced during thick film processing. Laser scribing or a diamond saw are used to cut substrates to the desired size. The nominal thickness commonly used is 0.025 inches.

The materials used to form the circuit pattern consist of conductor, dielectric, and resistor compositions. These thick film pastes have three main constituents :

Basic Thick Film Processing 27

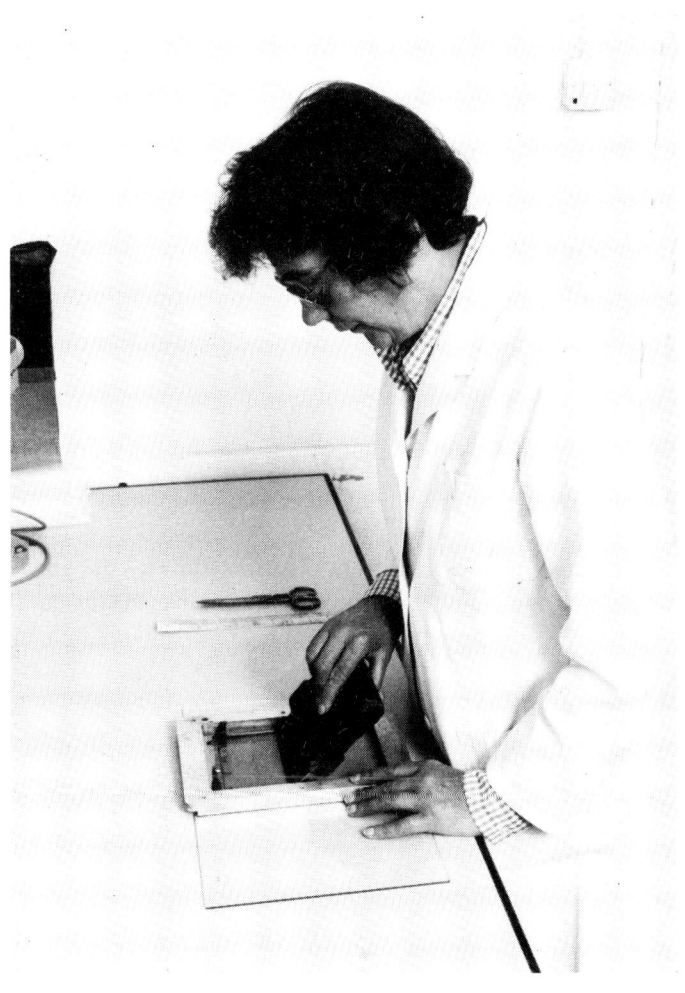

Figure 2.4 Coating a screen with emulsion.

(a) An organic thixotropic printing vehicle, which gives the paste the correct rheology for printing (the viscosity characteristic which is similar to that of emulsion paint).

(b) A glass frit to adhere the film to the substrate when fired and assist in sintering the particles of the filler to form a continuous film.

(c) A filler which is the material that will form the film on firing (e.g. gold powder in conductors, alumina powder in dielectrics).

The thick film manufacturer has a wide range of conductor materials to choose from depending on the application. Conductors based on gold and silver are widely used since they have good electrical characteristics and are not readily oxidised. Consequently they can be fired in an air atmosphere without degradation. Gold conductors are used for chip-and-wire hybrids where naked semiconductor devices are required to be wire-bonded into the circuit. Solder assembly with gold conductors is acceptable only if a gold-tin solder alloy is used (80% Au, 20% Sn, melting point 280°C). The leaching of gold by conventional lead-tin eutectic solder can be prevented by alloying the gold with palladium or platinum.

Palladium and platinum are also combined with silver to form palladium-silver and platinum-silver pastes which are used as low cost solderable conductors. The alloying elements help to inhibit silver migration and improve the solder leach resistance.

Copper thick film pastes have been in existence for some time but they have seen little use until recently. Thick film copper conductors have the lowest resistivities, typically 1.5 milliohms/square for a 15μm fired thickness print. The concept of ohms per square or sheet resistance is often used to characterise film resistivity. It is given by :

$$R_{sheet} = \frac{Bulk\ Resistivity}{Thickness}\ ohms/square$$

Basic Thick Film Processing

The value of any film resistor is then:

$$R = R_{sheet} \cdot \frac{L}{W} \text{ ohms}$$

where L is the length of the resistor and W its width. The same idea can also be applied to conductors, as in this case. In addition copper conductors possess good solder leach resistance and are not prone to migration. This material however is readily oxidised at elevated temperatures and hence must be fired in an inert atmosphere, usually nitrogen. The requirement for special processing has been the main obstacle preventing its more widespread use.

The newer generation of conductor pastes have either no glass frit or a low frit content. Copper oxide and cadmium oxide are typical materials which are used to chemically bond the film to the alumina substrate. This is known as reactive bonding.

Resistor systems are normally based on an oxide of ruthenium, chosen for its high temperature characteristics, good stability and TCR properties (temperature coefficient of resistance). This material when mixed with a glass frit and an inorganic binder forms the basic thick film resistor paste. The mix and shape of the particles that form the resistor on firing is a complicated topic and is covered in more detail in the chapter on thick film pastes. However suffice to say that by varying the proportion of conducting oxides within the insulating glass matrix a range of sheet resistivities results. The majority of resistor systems available fall into the range 1 ohm/sq to 10 Mohm/sq the standards being on a decade scale i.e. 10 ohm/sq, 100 ohm/sq etc.. Blending the basic members can be undertaken to achieve intermediate sheet resistivities. Specific resistor values are obtained by altering the aspect ratio of the resistor shape although in practice a post-firing tolerance of better than plus or minus 15% is difficult to achieve. Consequently resistors are designed to fire out lower than the desired value to allow for subsequent trimming.

TCR values of less than 100 ppm/deg C can be achieved from middle of the range members, typically 100 ohm/sq to 100 kohm/

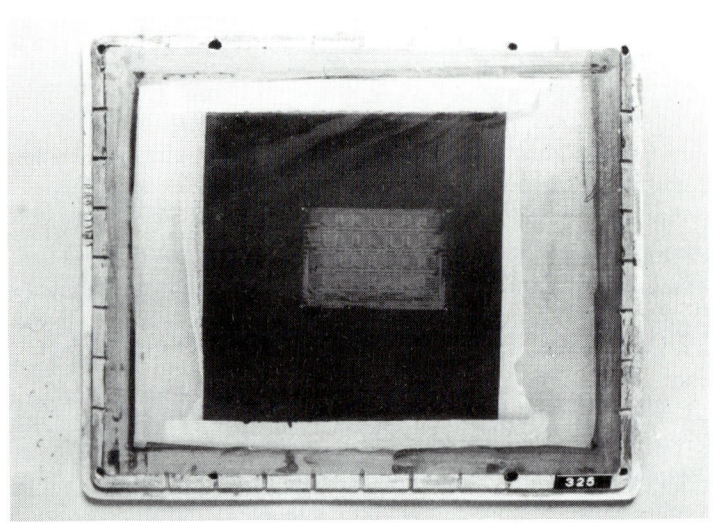

Figure 2.5 A completed thick film printing screen.

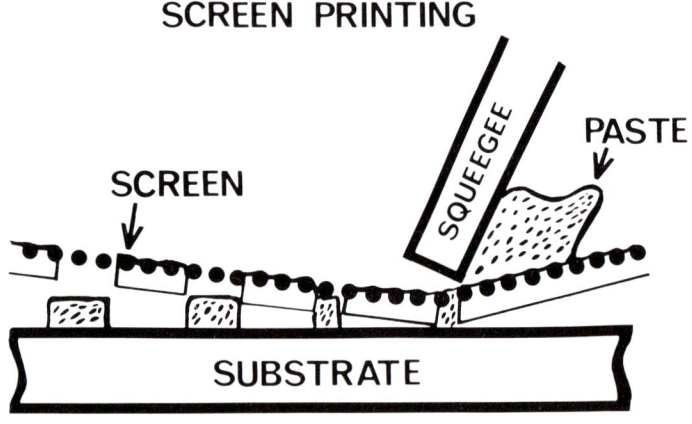

Figure 2.6 Schematic diagram of the printing process.

sq.. It is necessary to note that the conductor terminations play an important part in dictating the overall resistor characteristics since metal can diffuse into the resistor; this is known as the end effect. The stability of thick film resistors can be improved by coating the resistor with a low temperature firing glaze (usually 500 deg C).

Insulating dielectrics and glazes can be alumina loaded, depending on application. Dielectric constants can vary from typically four up to several thousand by the addition of high dielectric constant materials. Barium titanate ($BaTiO_3$) is used for capacitor applications. Low dielectric constant materials are used in multilayer applications to minimise capacitive coupling between conductors on adjacent layers. Via resolution for throughhole connections is typically 0.2 to 0.25 mm.

The substrate, screen and paste come together at the screen printer, as illustrated schematically in figure 2.6. A rubber squeegee blade traverses the top surface of the screen at a controlled speed, the pressure causing the thixotropic paste to flow through the pattern on the screen and hence onto the substrate. Screen printer set-up involves setting the correct snap-off (the distance between the screen frame and the substrate - typically 0.025 inches) together with the squeegee pressure and referencing to the previous layer. Snap-off and squeegee pressure affect to a large extent the thickness of the wet print. Too much pressure will cause spreading of the paste whilst too little will result in an uneven print. It is vitally important to control the thickness of resistor prints if consistent results are to be achieved. Wet print thickness can be measured using a light section microscope. A typical screen printer is shown in figure 2.7, and close up details in figure 2.8.

After printing the substrate is allowed to stand (to remove mesh marks), dried at 125 deg C to remove solvents, and then fired in air typically at 850 deg C. The temperature profile of the furnace is such that the organic printing vehicle is burnt off, and the film is sintered and adheres to the substrate. The peak firing temperature determines the amount of sintering and adhesion that will take place. Peak temperature must be controlled to within plus or minus 1 deg C again to ensure consistent results with resistors materials. Air flow within the furnace is important

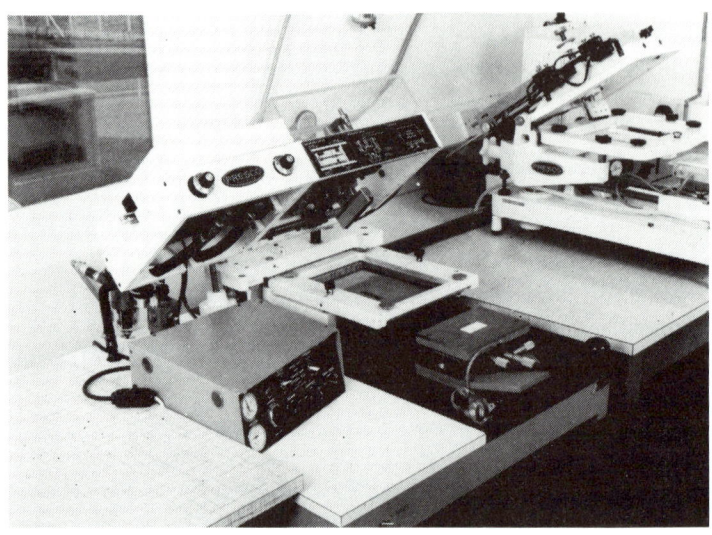

Figure 2.7   A typical thick film printer.

Figure 2.8   Detailed view of the printing head of a thick film printer.

Basic Thick Film Processing

to prevent burn-off gases interfering with conductors and resistors whilst they are in the peak temperature zone. A typical conveyor furnace is shown in figure 2.9.

The print-dry-fire sequence is repeated for each layer of the substrate, using resistive, dielectric or conductor pastes as appropriate. Resistors are printed last and follow a print dry print dry ..... fire sequence so that each resistor experiences the peak temperature in the furnace once only. Surfometer or Talysurf type instruments can be used to measure dried and fired film thickness.

Resistors are trimmed to the desired tolerance using an air-abrasive (sand blasting in miniature) or laser techniques. This is known as a passive trim and involves removing portions of the resistor whilst continuously monitoring its value. The resistor will increase in value until the required resistance is obtained. Trimming to a tolerance of 1% is readily achieved. Air abrasive trimming is slow and can cause damage to resistors other than the one intentionally being trimmed. The pulsed YAG laser is now the favoured approach for trimming. It is more suited to fast programable automatic trimming as the laser beam can be steered around the substrate and thus trim a number of resistors sequentially with no operator intervention. An automatic trimming station is shown in figure 2.10.

Active trimming is performed on assembled and tested substrates. In this case the resistor is not trimmed to a specific ohmic value. Instead the circuit is powered up and a circuit parameter such as gain, phase or frequency is monitored while the resistor is being trimmed. In this way circuit parameters can be defined very accurately despite the use of other low tolerance components, such as capacitors. The technique is useful for active filters, oscillators, precision amplifiers, digital to analogue and analogue to digital converters and similar circuits.

To complete the circuit prior to test, add-on components are assembled onto the substrate using epoxy die-attach or solder connect or a combination of both. Pre-packaged devices are soldered (e.g. chip carriers, SOT23 etc.). Gold or aluminium wire bonding is used to connect naked semiconductors onto the

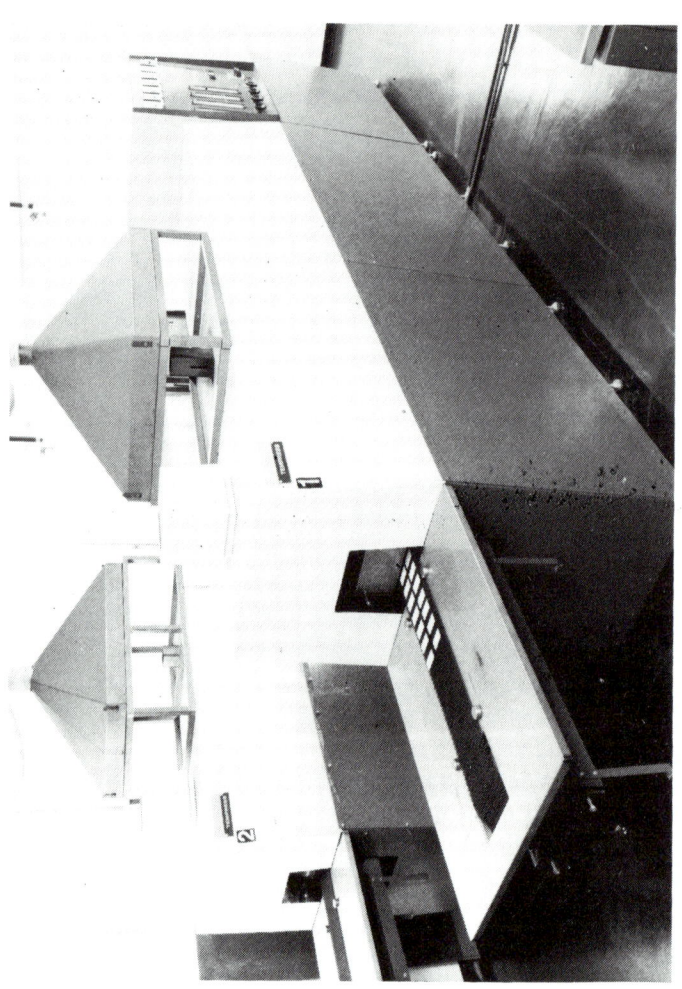

Figure 2.9 Thick film conveyor furnaces.

substrate - e.g. integrated circuits, diodes and transistors.

Packaging is normally semi-hermetic (e.g. plastic, potting epoxy etc.) or fully hermetic (e.g. metal welded can, solid sidewall package or glass sealed ceramic package.)

Final test and burn-in concludes the fabrication process.

PROBLEM AREAS

The hybrid industry has been faced with a number of challenging problems throughout its development and solutions to these problems continue to form the basis for discussion amongst those involved.

Design Layout

Human designers are extremely adept at working on two layers (e.g. a double sided PCB). However to obtain maximum benefit from thick film hybrid technology, a multilayer approach should be taken, particularly with the more complex circuits. It must be remembered that with hybrids the designer must configure not only the conductor layers, but dielectric and resistor layers as well. It has been found that most hybrid layout engineers find extreme difficulty in working with three or more conductor layers simultaneously. The use of CAD can help, but not all manufacturing facilities have this capability.

A partial solution has been found to this problem through the use of automated route-running programs although in general they do not provide an optimum layout from a performance point of view. However in many cases a suitable layout will result with associated savings in time and effort. If signal connections are constrained to two conductor layers with power supply and ground connections limited to conductor layers 3 and 4, a significant reduction in complexity of layout can be realised whilst still maintaining the advantages of a true multilayer technique.

Cumulative Yield

As the complexity level in hybrid circuits increases with

Figure 2.10   A typical thick film laser trimmer.

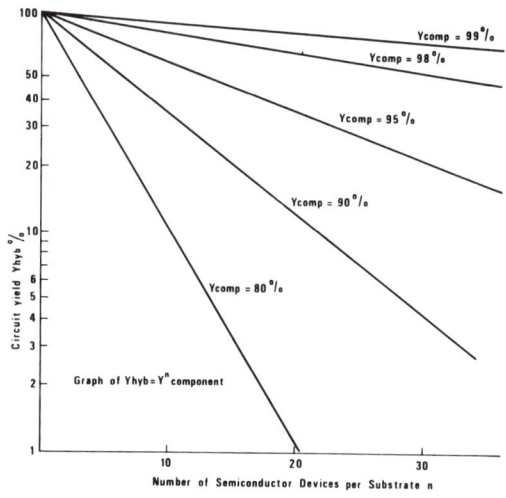

Figure 2.11   Yield characteristics for hybrid assembly.

Basic Thick Film Processing

improvements in layout technique and substrate fabrication methods, a difficulty arises regarding the final functional yield after assembly and wire-bonding of active add-on devices. IC's, transistors, diodes etc. are DC probe tested only at wafer level prior to dicing and inspection. The yield at this level, even after 100% visual examination, strongly affects the final yield of the hybrid. The overall hybrid yield is the cumulative yield of all active die used to make up the part. Clearly as the number of die per hybrid increases so the overall yield drops.

For example, suppose the die yield is 95% which at first sight would seem reasonable. If a hybrid contains two of these devices then the hybrid yield will be 90% (.95 x .95) which is again reasonable. This figure assumes of course that the substrate is perfect and that the assembly and bonding operations have inflicted no damage to the devices. If the circuit has five such devices then the yield will be 77% (.95 x .95 x .95 x .95 x .95) which is beginning to look less attractive. If a circuit comprises twenty-five such devices, which is a typical component count on a 2" by 1" substrate, then the yield will be only 28% which is far short of the original 95%. A general graph of overall yield against number of devices is given in figure 2.11.

In practice therefore repair will be necessary in order to increase the end yield. Testing at the chip level prior to assembly is possible; however it is not often undertaken due to practical difficulties in handling naked chips and the possibility of probe damage causing problems during wire bonding. Provided that the supplier of the devices has a good inspection system initial die yield can be as high as 99% which will result in an overall hybrid yield (on a 25 chip circuit) of 78%.

Where complex chips of a known low yield are to be incorporated into a circuit, tape bonding or chip carriers can be used which will allow functional pre-testing.

Materials Control

This is a difficult problem for the hybrid engineer. For the materials used in thick film production consideration must be given to the chemical, physical and metallurgical properties of the fired film. Materials can be evaluated with regard to printability, definition, spread, adhesion, resistance, stability, TCR and wire bondability. These characteristics are not necessarily independent and trade-offs need to be made. Effort spent on evaluation can put a significant overhead on the overall cost of the operation, particularly where many material interactions are taking place. In general it is beneficial to use a range of materials from one manufacturer, so that if subtle material interactions degrade the hybrid performance, then there is a clearly defined onus on the supplier to investigate the problem. Discussions with suppliers will allow such problems to be anticipated and prevent an incorrect choice of materials being used. Field failures are the responsibility of the hybrid manufacturer, not the paste supplier !

Thermal Management

Where power devices are necessarily incorporated (e.g. fast logic, power transistors etc.) an assessment must be made of the thermal behaviour of the hybrid module in its real life environment. This is an extremely difficult task, even when all the individual components and materials have been evaluated. The solution comes from the analysis of a multiple three dimensional heat source problem. Practical measurement is awkward due to the small sizes of components used. Various computer programs exist which can be used to provide estimates of thermal performance but these are by no means general, and they do not take into account transient thermal behaviour.

It is important to attempt, as far as possible, to quantify the junction temperatures and thermal impedances. It must be remembered that the reliability of semiconductor devices degrades rapidly when the junction temperature is over 125 deg C.

Testing

The increasing complexity of hybrid circuits can cause a number of difficulties for the test engineer. Cumulative yield has already been discussed.

In order to screen out bad devices, testing has to be performed

to diagnose faults. In many cases it is necessary to probe directly
into the hybrid circuit to trace signals, power supplies and so on,
with the attendant danger of some unintentional damage occuring
to otherwise functional parts of the circuit. This proves to be a
minor problem with small collections of analogue and MSI digital
devices. However if a bus structured system is involved manual
probing does not yield a great amount of information and fault
location becomes impossible. Logic analysers and in-circuit
testers can be used but the conventional 'bed-of-nails' approach
tends to be restricted due to lack of access to such a small area.

Clearly where microprocessor and memory type devices are
incorporated devices should ideally be pre-tested at temperature
to ensure their function is correct. TAB (Tape Automated Bonding)
and chip carriers have a part to play in the use of this type of
component. Effective partitioning of the circuit and the provision
of test points on previously un-used pins of the hybrid can go a long
way to easing the problems of the hybrid test engineer.

APPLICATIONS

Thick film hybrid technology is used in a wide spectrum of
applications with new areas being continually discovered. In
the automotive field typical examples include alternator voltage
regulators and engine control units. In telecommunications
receivers, transmitters, handsets, VHF and UHF filters can be
implemented in hybrid technology with benefits in size and
performance. Moving up in frequency to microwaves a number
of microstrip components together with antennae arrays can be
fabricated in this somewhat specialised area. Thick film microstrip
is a strong candidate for future developments in the field of microwave receivers for satillite TV systems. Finally in the military and
aerospace application areas the small size and weight requirements
have dictated the use of thick film for radar, thermal imaging
(man portable systems) and avionics processors to name but a few.

FUTURE DEVELOPMENTS

The use of base metal conductors (copper in particular), large
area substrates (metal cored) and surface mounted components such
as chip carriers will complement the well established chip-and-

wire thick film circuit to ensure a bright and expanding future for hybrid microelectronics. Thick film has yet to be fully exploited as a versatile interconnection technique.

© British Crown copyright 1983 (year of first publication). Published by permission of the Controller of Her Britannic Majesty's Stationery Office.

3
Thick Film Pastes and Substrates

B. WALTON

INTRODUCTION

Pastes and substrates are the basic building blocks of thick film circuits and in this area of hybrid microelectronics, progress has been rapid and many innovations introduced over the last decade or so. In general pastes are proprietary products manufactured by specialist companies supplying circuit manufacturers. Thus, although a wealth of technical data concerning the performance of these materials is available, comparatively little is published concerning their formulation. The relationships between formulation and properties will be reviewed in this chapter.

THICK FILM PASTES IN GENERAL

All thick film pastes used in hybrid microelectronics consist of a permanent inorganic content and a temporary organic vehicle. The inorganic material remains as a film after the firing process whilst the organic content is removed during drying and in the furnace used for firing. A typical firing profile is shown in figure 3.1. The inorganic content (largely metals and metal oxides) is present as a finely divided powder whilst the organic content is in the form of a viscous solution and the two are blended into a paint like consistency. The inorganic content varies considerably according to the type of paste (i.e. conductor, resistor, dielectric etc.) whilst the organic vehicles are basically similar for all pastes. The main function of the vehicle is to hold the inorganic material, which ultimately forms the desired final film, in suspension in a

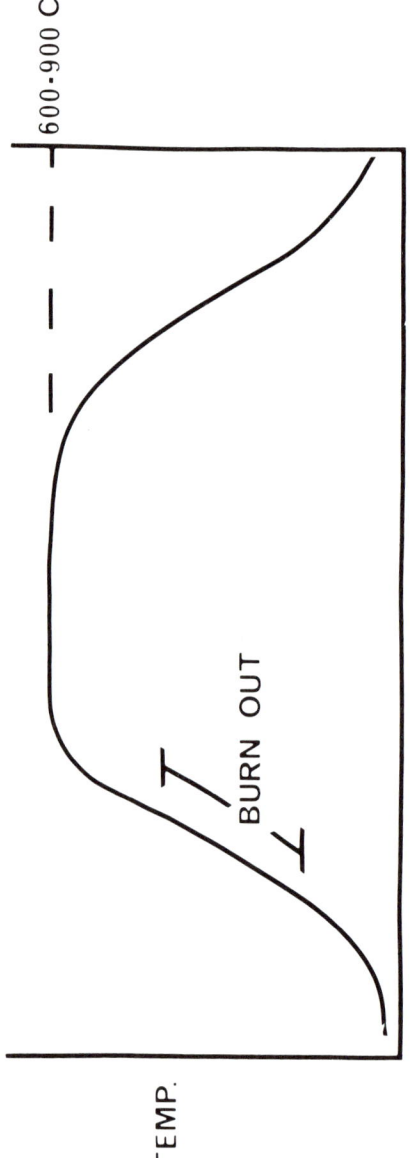

Figure 3.1 Typical thick film firing profile.

viscous liquid medium suitable for screen printing and hence film
pattern delineation. The vehicle normally consists of a blend of
differing grades of ethyl cellulose or other cellulose derivatives (1)
dissolved in a blend of solvents. This produces a viscous liquid
with flow characteristics (rheology) suitable for screen printing. The
rheology is rather similar to that of an ordinary paint: when the
liquid is forced to flow, i.e. subjected to a mechanical shearing
action such as takes place in brushing or screen printing, the
effective viscosity reduces as the rate of shear increases. On
removing the shearing forces the viscosity increases and flow tends
to cease. Thus in the printing operation as the squeegee blade
passes over the screen, the paste is rapidly sheared, the viscosity
falls and the paste passes easily through the fine mesh of the screen
onto the substrate. Shearing action then stops and the viscosity of
the paste regains its initial higher value and flow almost ceases.
Thus the paste does not flow out on the substrate and the shape of the
printed pattern is retained. In practice the rheology of the paste is
strongly influenced by the total loading of solid material and by the
particle size and particle size distribution of the powders used.
Minor additives, known as thixotropes, are used to produce the
exact rheology required. For fine line printing, pastes of relatively
high viscosity are used, whilst for high speed printing the opposite
is true. The blend of solvents used in the vehicle is adjusted to be
sufficiently volatile to allow drying to occur in a reasonable time,
but not so volatile as to give problems or rapid drying out on the
screen or in long term storage. After driving off the solvents the
dried film is held together by the remaining solid organic content,
known as the 'binder' which is burned off in the firing operation.

There are a number of special types of thick film pastes
available but the most important classes are conductors, resistors
and dielectrics, and these are discussed in the next sections.

CONDUCTORS

The primary requirements of thick film conductors are high
electrical conductivity and strong adhesion to the substrate. The
many secondary characteristics include :

> solderability
> suitability for bonding (ultrasonic, thermocompression

and eutectic)
printing capability
compatibility with other thick film materials
and substrates
processing conditions
ageing characteristics

Obviously cost is another significant and often dominant factor.

High conductivity is achieved firstly by selection of a metal or alloy with a high intrinsic bulk conductivity and secondly by maximizing densification during firing. Adhesion is achieved either by incorporation of a small amount of glass in the paste (2) (so called 'fritted' conductors) or else by the inclusion of certain metal oxides which will react chemically with the metal and underlying ceramic (these are known as reactively bonded conductors)(3,4). The inclusion of these additives impairs the conductivity and, in the case of fritted conductors, the structure of the metal at the interface must be to some extent porous to allow the glass to key mechanically the conductor to the substrate. This however also reduces conductivity. A section through a fired conductor is shown schematically in figure 3.2. Thus in general it is not possible to maximize adhesion and conductivity simultaneously and a compromise has to be made. In a similar way there are trade-offs to be made in secondary characteristics and consequently a very large number of formulations, each offering advantages for certain applications have become commercially available.

Conductors may be conveniently classified according to metal or alloy content, and the salient features of the main types are given in table 3.1.

RESISTORS

Resistive systems are perhaps the most sophisticated of the thick film materials. The main technical challenge in producing a thick film resistor series is that of obtaining a wide range of resistance value with a low temperature coefficient of resistance (TCR), coupled, of course, with stability over long periods of time at elevated temperatures or on electrical load. This is achieved

TABLE 3.1

Metal or Alloy	Principal Features
Silver	Very low cost. Highest conductivity, Silver ion migration and tarnishing problems. Seldom used in microcircuits. Migration occurs through glass as well as between conductors.
Palladium silver	Low cost. Palladium inhibits migration but lowers conductivity.
Platinum silver	Low cost. Smaller quantity of platinum replaces palladium in the binary alloy.
Gold	Very expensive. Highly conductive and chemically inert. Reliable bonds to gold wire. High solubility in common solders makes soldering difficult.
Palladium Gold	Compared with gold the solubility in solder is reduced but conductivity impaired.
Platinum Gold	Reliable solderable alternative to gold but not so conductive. Very expensive.
Copper	Fairly low cost. High conductivity. Has to be fired in neutral or reducing atmospheres to achieve excellent solderability.
Nickel	V. low cost. Can be fired in air but not solderable from the furnace.

Figure 3.2   Cross section of a typical thick film conductor.

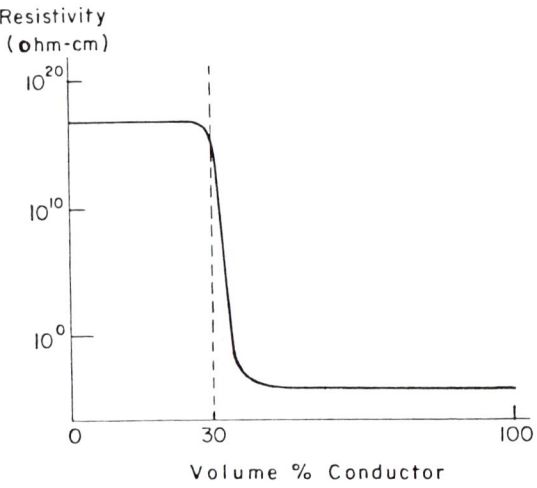

Figure 3.3   Classical conductor-insulator blending curve.

by arranging for a conducting phase to be dispersed in a non-conducting phase (usually glass) in the fired film.

The choice of materials for the conducting phase is very restricted. Metals have very low resistivities and generally large positive temperature coefficients. Semiconductors have higher resistivities but generally large negative TCRs. Also, the conducting phase has to be chemically stable when exposed to the furnace atmosphere at high temperature and when in contact with molten glass. Ruthenium dioxide and related mixed oxides such as bismuth ruthenate are unique compounds in that they combine moderate electrical conductivity with a near-zero temperature coefficient of resistance. These oxides are also stable in air at typical thick film firing temperatures (about 800°C). They are therefore the basis of the conducting phase in most successful thick film resistor systems used today, sometimes in combination with precious metals (5,6). A range of resistance value is obtained basically by varying the concentration of conducting oxide in the glass.

However this cannot be achieved in a controllable manner simply by dispersing the conducting phase in the glass in a random fashion since this results in a very abrupt change in conductivity as the conductor:glass ratio is varied. Such a typical conductor to glass blending diagram is shown in figure 3.3, and in order to obtain a very wide range of resistance value (up to 6 or 7 orders of magnitude) with low TCR (typically 200 ppm/°C or less) the geometrical distribution of the conducting phase in the glass matrix has to be non-random and carefully controlled. A typical resulting dispersion of conductive and insulating material is shown in figure 3.4 (7).

The materials used are expensive and the manufacturing procedures complex. Hence precious metal resistor pastes are expensive. Some attempts have been made to introduce base metal resistor systems but most of these have to be fired in neutral or reducing atmosphere and have rather inferior characteristics (8).

DIELECTRICS

Dielectric films are used in hybrid circuits to provide insulation between successive layers of conductor, as capacitor

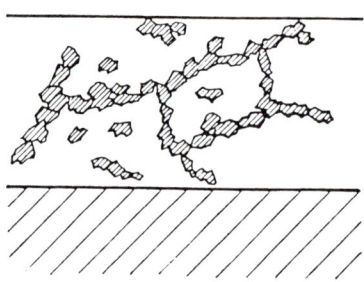

Figure 3.4   Conductive dispersion in a typical thick film resistor used to avoid abrupt changes in blending characteristics.

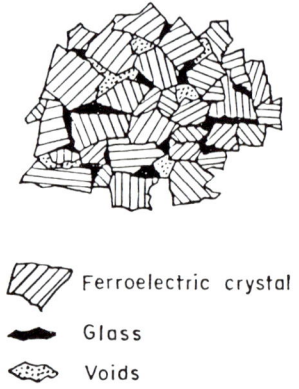

▨ Ferroelectric crystal
◆ Glass
░ Voids

Figure 3.5   Structure of a high permittivity thick film dielectric.

dielectrics, and as a final protective coating. The properties and composition vary accordingly.

For multilayer work, as well as providing insulation it is desirable that the dielectric has a low permittivity to minimise capacitive coupling between layers, and for use in high frequency circuits a reasonably low dielectric loss. Perhaps the most important feature of modern multilayer dielectrics, however, is the fact that these materials can be re-fired several times without further flow occurring. In the early days of multilayer dielectric development the undesirable re-flow of fired layers which occurred when circuits were passed through the furnace to fire successive layers could only be avoided by firing at progressively lower temperatures and the composition of each layer had to be adjusted to make this possible. Modern multilayer dielectrics can be re-fired at the same temperature several times without further flow so that fine details of shape, particularly through holes, (or vias, as they are usually called) are not lost. This is generally achieved by the use of glass-ceramics which are present in the paste initially as glassy particles. They readily soften and sinter on heating, but then crystallise on cooling and cannot be softened by reheating to the same temperature (9). An alternative technique is to arrange for the dielectric to consist of a glass mixed with crystals such as silica or alumina which dissolve in the glass thus raising the softening temperature of the composite (10). The glasses are typically based on lead boro-alumino-silicate systems and contain numerous minor additives to achieve the desired adhesion, flow, hermeticity etc. properties. Again, any particular blend is a compromise between a number of conflicting requirements.

Capacitor dielectrics should have a high electric strength, a high permittivity and low loss. These three properties are not readily achieved in a single material. Low loss dielectrics of high electric strength with low permittivities are available for low value capacitors. High permittivity materials on the other hand all have high losses and may have relatively poor breakdown strengths. A dielectric with a permittivity of 1000 will typically have a loss factor of a few percent. Many high permittivity dielectrics consist of ferroelectric crystals bound together with a small amount of glass. In order to maximize the permittivity, the glass content must be kept to a minimum and this tends to result in a somewhat porous structure, as shown in figure 3.5. The effect of the air voids is highly significant and reduces the measured permittivity considerably and gives rise to poor electrical strength. Better results can be obtained with special glass ceramic materials which yield a ferroelectric phase when they devitrify (11, 12).

Overglazes consist simply of glasses which have a reasonably low softening point (to avoid a final high temperature firing), good chemical durability, and a coefficient of expansion which is a reasonable match for the substrate.

It is important in the formulation of thick film dielectric materials to avoid the use of any material that might be mobile in the dielectric, particularly when under the influence of electric fields.

SUBSTRATES

From the users point of view, substrates provide essentially a passive mechanical platform for the circuit elements. This is not strictly true because there is usually some degree of chemical reaction between the substrate and thick film layers. This may markedly influence the adhesion between substrates and conductors and differing minor chemical constituents in ceramic substrates can therefore sometimes be important in this respect. For example, because of differences in availability of raw materials, alumina substrates of European origin may vary slightly in composition from those made in the U.S.A. and this may require modification to matching conductor systems (13).

Substrates are usually made of debased alumina, that is, more or less pure alumina oxide crystals bound together with a small proportion of glass with a lower softening temperature than the alumina. In general, the higher the glass content the lower the firing temperature used to manufacture the substrate and hence the lower the cost. However, excessive glass content lowers the thermal conductivity and mechanical strength of the alumina. Beryllia, which has a higher thermal conductivity than alumina, has been used but only in very special circumstances.

The main substrate problems encountered by users are concerned with dimensional tolerances: size, flatness, and surface smoothness. The difficulties of maintaining accurate dimensions and flatness result directly from the enormous shrinkage which occurs when the ceramic is first sintered at high temperature by the ceramic manufacturer. This shrinkage may be as much as 20% in linear dimensions and is not highly reproducible. Finished dimensional tolerances (as fired) are therefore typically no better than 1% and often wider. The advent of CO_2 laser scribing has meant that more accurate lengths and widths can be obtained if needed. Processing of flat ceramic substrates has greatly improved over the last fifteen years and substrates with bow of less than 0.001" per inch are obtainable without recourse to surface grinding. Surface finish has also been improved and is now more than adequate for thick film purposes.

The use of vitreous enamelled steel substrates (or in American terminology porcelain enamelled steel) has recently attracted a lot of attention. These substrates are low carbon steel sheets coated with glass. Because the glasses used begin to soften at temperatures above about 600°C, special low temperature firing thick film pastes have been developed for use with steel substrates. The principal advantages of steel substrates are mechanical robustness, availability in large sizes at low cost compared with ceramic and high thermal conductivity compared with printed circuit boards. However, at the present stage of development, the electrical characteristics of thick film elements fired on enamelled steel are inferior to those obtainable with conventional thick film pastes used with alumina (14).

FUTURE TRENDS

Future developments in thick film materials and substrates can be expected to reflect economic pressure for cost reduction and the need to meet the technical requirements generated by the trend in microelectronics towards ever higher packing densities and switching speeds.

The most obvious scope for cost reduction in materials is in the replacement of precious metals by cheaper, base metal alternatives. Nitrogen-firing copper conductors and matching dielectrics seem destined to improve and find a place in future multilayer circuitry, particularly large digital systems (15). Glazed steel or similar dielectric coated metal substrates with improved characteristics are likely to be used in situations where large area boards are required. For new generations of dense, high speed circuitry, interconnection networks based on new thick film materials applied to metallic substrates with very high thermal conductivity may well emerge.

Although not strictly based on conventional ceramic thick film technology, the use of screen printed polymer materials doped with silver (for conductors) or carbon (for resistors) may well be used in large quantities for consumer equipment in the future. The low cost of the materials plus the ability to employ very inexpensive substrate materials is very attractive in non-demanding applications (16).

4
Add on Components and Attachment Methods

J. R. POLDEN

INTRODUCTION

The components and mechanisms of component attachment are the devices and skills that turn a film resistor network into a hybrid and are therefore fundamental to the successful production of hybrid circuits. Although the processes are well understood and fully documented, the need for considerable process control must not be overlooked and careful monitoring is necessary to acheive high yield and a quality product.

A typical thick film circuit is shown in figure 4.1 and illustrates many of the common components and attachment methods. The substrate is 96% alumina, the most commonly used material. The film conductors and resistors can be clearly seen as the grey and black areas respectively. The capacitors are reflow soldered into position; the unencapsulated integrated circuits are glued in place and are electrically connected by wire bonding. The connections between the substrate and package pins are also made by wire bonding. There are therefore two distinct aspects to component attachment, the mechanical method by which the component is fixed to the substrate and the technique of electrical connection to the remainder of the circuit. In the case of reflow soldering the same process is used for both, whereas for the unencapsulated integrated circuit these two aspects are distinct.

THE SUBSTRATES AND METALLIZATION

The vast majority of thick film circuits use 96% alumina as

Figure 4.1 Typical metal packaged hermetic hybrid circuit demonstrating some of the connection techniques.

the substrate material and either palladium silver or gold as the conductor. This combination can withstand high temperatures and therefore can be soldered (in the case of gold using solders that are not based on tin or that already contain significant amounts of gold), and can have wires bonded to them.

There are however many possible alternatives and thin film hybrids use either 99% alumina or glass as the substrate and metallic gold as the conductor. The gold layer is often less than 1 μm in thickness and therefore any form of soldering is difficult. Recent innovations in thick film technology, in the drive for lower cost substrates, have included the use of plastics, such as polyimide and mylar, and dielectrically coated metals (typically porcelain enamelled steel). Each of these has its own characteristics that affect the methods by which attachment may be made.

THE COMPONENTS

By far the most common add-on components used in hybrids are capacitors and semiconductors. However most components and component styles have at some time or other been attached to hybrid circuits and there are many instances of inductors, transformers and occasionally switches being incorporated on the substrates. Essentially if there is a demand to attach a component to a hybrid, a way will be found. The large demand for semiconductors and capacitors does mean that these components are available in styles amenable for attachment to hybrid circuits. Electrically the components are similar to more conventional package styles, with the exception that unencapsulated integrated circuits have not undergone full parametric testing. However the range, stability and dielectric characteristics of the capacitors most frequently employed in hybrid technology are rather limited and will therefore be reviewed here.

Capacitors

Capacitors packaged in a style suitable for attachment to hybrid circuits are available in value from a few picofarads right up to about a hundred microfarads. Not surprisingly, in order to obtain high capacitive density there are some quite severe restrictions over the secondary parameters, such as stability, temperature coefficient and dissipation factor. Up to about 0.5 µF the usual dielectric material is ceramic and a table of comparative values are given in table 4.1. Thus, for signal processing and similar high stability applications it is essential to use the NPO dielectric and hence the range of values available is restricted to about 20 nF. The devices are in the form of a rectangular block, or chip, with metallization at the two ends. The sizes vary from .05" x .04" in plan area up to about 0.25" x 0.25" depending on the value and dielectric. Some typical components are shown assembled in figure 4.1 and the mid-range dimensions are in the region of 0.12" x 0.10" and 0.90" x 0.70".

Tantalum electrolytic capacitors are also available in various chip forms with metallization either on two ends of the component or sometimes, one of the electrodes is a small wire protruding from the centre of the device. Typical values are from 0.5 µF with up to a few tens of volts breakdown, to 100 µF at 3 to 5 volts breakdown. Again sizes are in the region of 0.15" by .08" in the mid values.

Semiconductors

Semiconductor devices are available in a variety of styles in addition to the standard 0.1" dual-in-line format. The most common of these styles are the SO package, which is basically a miniature version of the standard DIL package with the leads preformed for mounting to hybrid circuits (transistors are also available in this style) and the chip carrier. In both these techniques the semiconductor circuit can be parametrically tested prior to assembly and hence there is a degree of assurance that the final yield will be high.

Most devices are also available as the basic dice, either as the sawn slice mounted on a tacky plastic foil or in waffle trays. The slice will have been subjected to a d.c. test prior to isolating the individual devices. The faulty circuits are marked during testing and would not be shipped from the manufacturer when supplied in waffle trays. They would however be included when the slice is mounted on tacky foil. An important advantage of using the tacky foil technique is that the history of each device is known and hence adjacent and nominally matching devices on the slice can be

TABLE 4.1

DIELECTRIC TYPE	ULTRA STABLE NPO	STABLE HIGH K X7R	STABLE HIGH K BX
CAPACITANCE RANGE	1 pF to 0.027 µF	100 pF to 1 µF	100 pF to 0.47 µF
TOLERANCES	$\pm$.25pF or $\pm$.25%, $\pm$.5pF or $\pm$.5% $\pm$1%, $\pm$2%, $\pm$5%, $\pm$10% dependent upon Capacitance Value	$\pm$5%, $\pm$10%, $\pm$20%	$\pm$5%, $\pm$10%, $\pm$20%
TEMPERATURE RANGE	$-55°C$ to $+125°C$	$-55°C$ to $+125°C$	$-55°C$ to $+125°C$
TEMPERATURE COEFFICIENT /CHARACTERISTIC	$C < 20$pF $^{+120}_{-40}$ ppm/°C $C > 20$pF ± 30 ppm/°C	$\pm 15\%$ at 0 V d.c. over rated temperature range	$\pm 15\%$ at 0 V d.c. $+15, -25\%$ at rated voltage, over rated temperature range
VOLTAGE RATINGS	50/100 V d.c.	50/100 V d.c.	50 V d.c.
DISSIPATION FACTOR	.001	.025	.025
INSULATION RESISTANCE	100,000 MΩ or 1000 MΩµF whichever is less, at 25°C. 10,000 MΩ or 100 MΩµF whichever is less, at 125°C.		
PROOF VOLTAGE	2.5 x Rated Voltage for 5 seconds, 50 mA maximum.		

selected for certain applications. Similarly some of the circuits can be packaged and tested so as to obtain a guide to the likely performance of the remaining devices on the slice.

Problems can arise when using unencapsulated semiconductor components. Each manufacturer may use a slightly different layout for the same circuit and the physical outline may change from time to time as the manufacturer improves or alters his production process. Care must also be taken to ensure that the application does not rely on a critical parameter of the semiconductor device since this will not generally be determined until the device is mounted.

Typical device sizes range from 0.015" x 0.015" for a low power transistor, to 0.05" x 0.05" for a small scale integrated circuit, to about 0.2" x 0.2" for a complex large scale device.

ATTACHMENT

As already mentioned the attachment has two purposes, firstly to anchor the component firmly in place and secondly to provide electrical connection. These may sometimes be achieved by one process or in some cases separate processes. The three main attachment methods are:

(a) Welding, two metallic components are fused at high temperature or high stress (e.g. ultrasonic "scrubbing").
(b) Soldering, an intermediate metallic material is used which, when molten, dissolves the surface of the metallic components to be joined.
(c) Adhesive bonding, an intermediate organic material is used at relatively low temperatures. The glue may be loaded with either a conductive material (typically silver) to provide electrical connection or a thermally conducting material to aid heat dissipation.

Capacitor Attachment

The most commonly used capacitors are the ceramic multilayer chip devices as discussed in the previous section. These have metallized terminations at each end and the usual method of attachment is by reflow soldering. Typically a paste consisting of finely divided solder balls and a fluxing agent is screen printed or stencilled onto the substrate. The component is then placed in the solder paste whose "tackiness" is adequate to prevent minor bumps and knocks dislodging the component. The entire assembly is then heated to melt the solder and hence join the component to the substrate. This process is carried out just once for all the components that are to be reflowed. The heating needs to be carried out with care to prevent excessive dissolution of the film metallization in the solder (known as leaching) and also oxidization of the solder. Typical heating methods are a belt moving over a series of hotplates and dipping into the vapour of a fluid whose boiling point is above the melting point of the solder. This latter method is called vapour phase heating and has the important characteristics of heating the whole assembly uniformly and also of providing a non oxidizing atmosphere. The joints made by reflow soldering are known to be reliable and easy to inspect, but care must be taken to avoid small balls of solder forming and becoming lodged under components.

A variation of this process is to bond the chip capacitor to the substrate with a non-conducting adhesive and then to dip the substrate into a bath of molten solder after suitable fluxing. Alternatively the substrate can be passed through a wave soldering machine and treated in an analogous manner to a printed wiring board. In all soldering methods it is important that the flux is adequately cleaned off the substrate. Tantalum (electrolytic) capacitors can be treated in a similar way to ceramic capacitors.

If it is desired to avoid the use of solder, for instance when the conductor is gold, then it is possible to use conducting adhesive for the capacitor attachment. The adhesive is either screen printed or transferred from a reservoir to the substrate by means of a shaped tool. The adhesive is then cured at typically 150°C for one hour. This method has the advantage of not requiring any post attachment cleaning but some doubts have been expressed concerning the effect of long term outgassing of the adhesive, although a very large number of components are in field service using this method.

Occasionally capacitors are fixed in position with a non conducting adhesive at the centre of the component and then the electrical connection is made by wire bonding in a manner similar to the method to be described for integrated circuits. The important characteristic of this technique is that it allows any relative movement caused by thermal expansion mismatch to be absorbed in the wire bonds.

Lead Frame Attachment

In the manufacture of industrial and consumer grade hybrids and also the integral substrate package hermetic grade hybrid, it is necessary to attach a lead frame. The frame is invariably clipped onto the substrate in either a dual in line or single in line format. The permanent attachment is then achieved by dipping the frame and substrate edge into a pot of molten solder. A more recent variation of this process is to include a small stick of solder in the lead frame and the whole assembly would then be heated in a vapour phase heating system. The solder then flows down the clip and onto the substrate to form the joint. Generally this latter technique would be carried out at the same time as the other components are reflowed, whilst the former would be a second soldering operation.

Standard Integrated Circuit Package Attachment

Integrated circuits in standard packages would only be used as a last resort when one of the other package styles is unacceptable for reasons of price, quality or availability. The leads of the package are formed so that it is possible to stand the device on the substrate and to use reflow soldering in the same manner as the chip capacitors are attached. It is difficult to automate the lead forming and placement of the standard packages.

Miniature Packaged Semiconductors

There are two basic styles of miniature packages, the "SO" and the chip carrier. The SO package, is basically a plastic dual in line package with the leads on 0.05" centres and preformed for surface mounting. The chip carrier is essentially the cavity from a standard ceramic dual in line package with lead outs formed by tracking down the side and underneath the package. The advantages of the chip carrier are that it is hermetic, can have up to 84 leads without a significant space penalty and can be used to package and test die that are otherwise not obtainable in surface mounting styles. As it is a ceramic construction, it has an excellent thermal match to the alumina substrate, and therefore soldered joints are not stressed when the ambient temperature alters. Both these package styles are reflow soldered in position.

Die Attachment

The basic die is obviously the smallest form of semiconductor available and hence gives rise to the smallest overall hybrid circuit. However unless long production runs are envisaged or the ultimate in miniaturisation is required, then it is to be a relatively costly way of putting semiconductors devices onto a substrate since the die tend to be no cheaper than the packaged components. The mechanical attachment of the die is usually by means of conductive epoxy resin, although gold-silicon eutectic attachment is also possible. The electrical connections are made by wire bonding.

Wire Bonding

There are three methods of wire bonding available, aluminium ultrasonic, gold thermocompression and gold thermosonic. The aluminium ultrasonic technique is a wedge-wedge bond as illustrated in figure 4.2. On the first bond, the bonding tool descends and traps the wire between the tool and the integrated circuit. Ultrasonic energy is then applied and the scrubbing action causes the aluminium to flow and join the two surfaces. The tool is then moved to the bonding site on the substrate and the process repeated. After the second bond has been made, the heel of the tool is used to dent the wire and on moving away from the joint the wire snaps. The cycle can now be repeated. The major limitations for hybrid technology of this method of bonding are that the tool can only move in a straight line once the first bond has been made and that when used with gold thick film conductors the joint ages and weakens fairly rapidly. Thermocompression bonding uses gold wire with the cycle illustrated in figure 4.3. Here the first bond is made by forcing the capillary tool onto a ball formed on the end of the wire with a background heat of about 300°C. The second joint is then formed as a wedge bond in a similar manner. The capillary is shaped to weaken the joint in the centre and hence the wire breaks at this point when the tool moves away. The ball is then formed at the end of the wire, usually by causing a spark to jump between the end of the wire and an electrode. This method has the

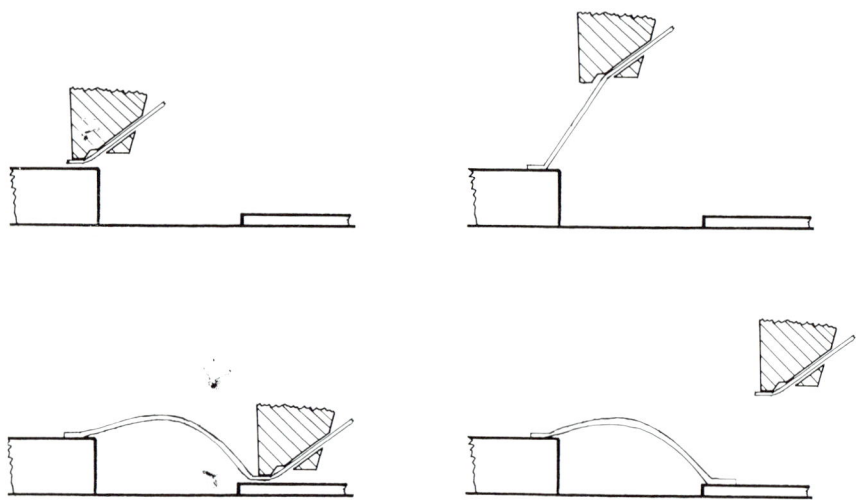

Figure 4.2 Aluminium wedge-wedge bonding.

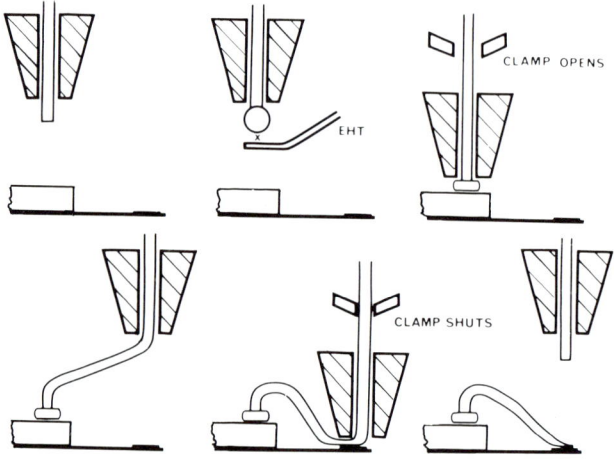

Figure 4.3 Gold thermocompression or thermosonic ball-wedge bonding.

advantage that the tool can move in any direction after the first bond has been made, thus allowing a relatively wide search area, but the high background heat is a distinct disadvantage. Thermosonic bonding is a combination of both methods where the gold wire ball-wedge bond technique is aided by ultrasonic energy. The background heat can then be reduced to about $100^{\circ}C$ - a temperature that does not cause any damage. As there may be several devices to be bonded the time taken for the bonding can be significant. Thus the lower temperature is quite an important advantage.

The thermosonic bonding technique is the one used for hybrid circuits although aluminium bonding to silver palladium conductors is also used when it is desired to eliminate the gold conductor for reasons of cost and yet still employ unencapsulated integrated circuits.

CONCLUSIONS

It can be seen that there are a variety of techniques available to the hybrid designer and it is worth bearing the following points in mind.

(a) There is always a choice of processes to attach any component. Those described are the most commonly used and are well understood.
(b) The choice of packaging that the component will be used in is the first consideration.
(c) The order in which the processes is carried out is important and must be carefully thought through. There is usually more than one order and one set of processes.
(d) The processes described are reliable if - and only if - satisfactory quality control procedures are built in.

5
Thin Film Processing

H. T. LAW

INTRODUCTION

A very large fraction of the output of the hybrid microcircuit manufacturing industry today is produced by the thick film process. In value of sales thick film hybrid integrated circuits lead by a factor of about 5:1. In terms of numbers of circuits manufactured this ratio is even larger due to the higher unit cost of the thin film hybrid microcircuit. Nevertheless the thin film process has its place in the complete spectrum of technologies which become involved in the hybrid manufacture and sufficient time has now gone by to demonstrate conclusively that this place will be retained because of certain specific advantages which the thin film technology has to offer. It is one of the purposes of this chapter to highlight the differences between the two technologies and to illustrate those situations in which the thin film process may be the one to choose.

Twenty or so years ago the concept of thin film circuit included, at least as an objective, the realisation of the entire circuit function in thin films. All of the necessary electronic components – resistors, capacitors, inductors, diodes and transistors may be generated by thin film technology and their integration into an "all film" electronic circuit is a completely valid concept of great appeal. In the event this objective has had to be gradually abandoned, the elegant has given way to the expedient and in the present-day hybrid thin film circuit it is usual to find only the resistor and the conductor pattern being realised in film form with extensive use being made of added discrete components, in a manner similar to that in thick film manufacturing processes.

PREPARATION OF THIN FILMS

Historically many methods have been used to deposit thin films of metals and dielectrics upon a number of possible substrates. These include :

(for metal films)
- a. electro-deposition
- b. chemical precipitation

(for metal or dielectric films)
- c. thermal reaction at heated surfaces (pyrolysis)
- d. vacuum evaporation
- e. cathodic sputtering

The majority of deposited films today are produced by sputtering, usually using radio-frequency electric fields. It is increasingly common to find "magnetron" sources used; these apply high magnetic fields in the region of the cathode to increase plasma density and thus the rate of sputtering. Vacuum evaporation is still used occasionally in the production of specialised resistive films.

Although it is a relatively simple matter to produce a deposited film, the achievement of reliable and reproduceable electric characteristics requires extremely careful process control. The deposition technology is very demanding and consistently high quality can be attained only if careful control of the many factors influencing the process can be achieved. The physics and the chemistry of the deposition process is complex. The electrical properties of the metal in thin film form may differ very substantially from those of the bulk material. As an illustration of some of the chemical factors which bring this about it should be recognised that :

- a. In the case of metal alloys deposited by vacuum evaporation the composition of the deposited alloy may differ from that of the source due to the different vapour pressures of the elementary metals which make up the alloy. (This problem is to a large extent circumvented by the use of sputtering as a method of deposition).

b. Since the rate of deposition of the film is small, considerable chemical reaction takes place between the growing film and the ambient gases in the deposition chamber.

The physics of the deposited film is also unusual in consequence of the fact that the inter-atomic forces between the metal atoms are much larger than the forces between the atoms and the material of the substrate. The magnitude of these forces can be determined by measurements of the latent heat of evaporation and it is found that (for nickel-chromium) the latent heat from the bulk material is some seven times greater than the latent heat of evaporation of a thin film from the surface of a glass substrate. Because of this preferential affinity, very thin films tend to agglomerate into "islands" and the conduction process is a combination of electronic conduction through these islands (modified by surface effects, since the electron mean free path is comparable to the thickness) and by tunnelling between the islands. The temperature co-efficients of these two conduction processes are opposite in sign so that we have the opportunity to obtain temperature co-efficients near to zero if the "thickness" (more correctly this should be described as the surface density) is chosen appropriately.

The materials used as resistor films are most commonly nickel-chromium and tantalum. The latter has the advantage that it forms a strongly adherent oxide layer with good dielectric properties which may be used to form thin film capacitors. Tantalum resistors may be adjusted to value (trimmed) by controlled oxidation in an anodising bath.

The production of consistently good quality resistor films (high stability and small temperature co-efficient of resistance) calls for very careful control of every aspect of the deposition process. The film properties are considerably affected by

substrate material,
substrate pre-cleaning and pre-treatment,
substrate temperature during deposition,
source composition,
deposition rate,
angle of incidence of evaporation (or sputtered) atoms,
ambient atmosphere in deposition chamber,

to list only the more significant parameters. Variation in the properties from week to week and from batch to batch, is the central problem in the production of improved thin film resistor performance.

The control problems are so daunting that an increasing number of manufacturers no longer deposit their thin films "in-house", but instead purchase "uncommitted" (pre-deposited) substrates from specialist manufacturers. As the description of thin film processing methods will show this separation of the deposition from subsequent process steps involves no loss of versatility, since it is not necessary, or usual, to modify the resistor film properties to suit particular required component values. The sheet resistivity of the film, for example, is constant and the resistance value and rating of the resistor is determined solely by its geometry.

For conductor films, pure gold has been used extensively although the rising cost of this metal has led to increased interest in other materials. Gold has the very important advantage that it is compatible with the thermal compression bonding process which uses gold wire to form the connections between active semiconductor devices and the film circuit, between the circuit and the package pins and, in some cases, for crossovers. For reasons which are clarified later it is not usual, nowadays, to find a crossover technology based on thin film processes and "wired" crossovers, using some form of microbonding technique, are used instead.

THE THIN FILM COATED SUBSTRATE

Figure 5.1 shows the structure of a typical plate of uncommitted precoated substrate material on which the later discussion on processing will be based. There are many possible variations in processing method; the one described is representative.

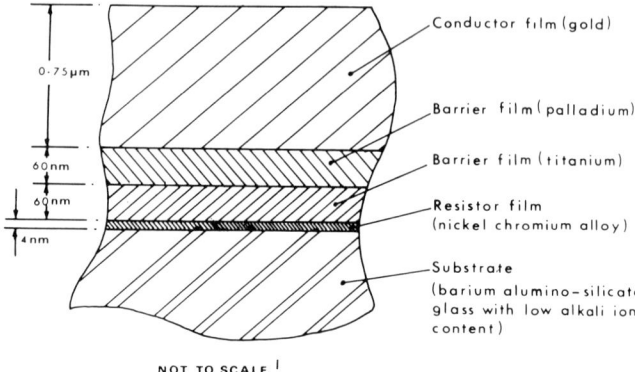

Figure 5.1   Build up of a typical "uncommitted" thin film coated substrate.

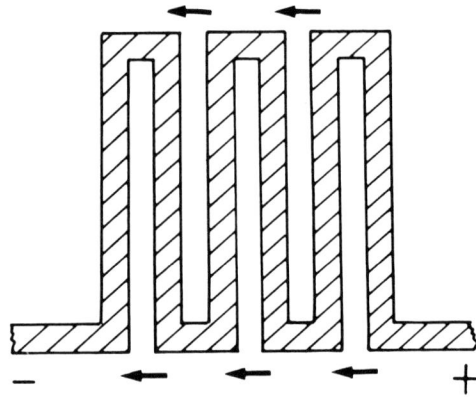

Figure 5.2   Illustration of the movement of alkali ions in the substrate under the influence of the electrical field existing between the meanders of a thin film resistor.

The substrate material is a glass having a very low free alkali ion content. Alkali ions in the glass have a finite mobility, so that when a resistor is under load the ions will migrate under the influence of the electric field which exists between adjacent "meanders" of the resistor, as shown in figure 5.2, and will react chemically with the thin film material. The direction of movement is shown in the diagram. This reaction will irreversibly increase the resistance, perhaps putting the resistor outside tolerance, and may eventually lead to total failure by open circuit. Since the mobility increases with temperature the effects are more serious in a circuit operated at high substrate temperature, for example at maximum rating in the maximum ambient temperature permitted. Barium alumino-silicate glass, type 7059, manufactured by the Corning Glass Company, Corning, New York, U.S.A. is suitable for the substrate material. (1)

After thorough chemical cleaning the bare substrate plates, 50.8 mm x 50.8 mm, are loaded into the deposition chamber in clean room conditions. Further cleaning of the substrate surface by ion bombardment takes place following pump-down, after which the resistor film (nickel-chromium alloy) is deposited. As previously stated the very thin resistor film is agglomerated, only partially covering the surface. The surface mass density of the metal is such that, if considered as a uniform "film" it would have a thickness of 4nm (40 $\overset{\circ}{A}$). This film has a sheet resistivity of 300 ohm/square and a temperature coefficient of resistance which crosses through zero at about $40^\circ C$.

Without breaking the vacuum other sources are used to deposit films of titanium and palladium, each 60 nm in thickness, and of gold, 0.75 μm in thickness. The titanium acts as a barrier film to prevent diffusion of the resistor film materials, especially the chromium, to the surface of the gold conductor film. At the temperature used in the thermocompression bonding process, and other assembly steps, sufficient chromium may otherwise diffuse to the surface, oxidise and prevent the formation of satisfactory thermocompression bonds.

The gold film has a resistivity of about 0.1 ohm/square. This may be inconveniently high. In a complex circuit involving long conductor paths a resistance of a few tens of ohms may be common.

It is not usual to vacuum deposit thicker gold films in order to reduce the resistivity, due to the high price of the metal and the fact that considerable mechanical stresses develop between thick gold overlays and the underlying films which may lead to separation of the gold. Selective electroplating is a more satisfactory way of reducing conductor track resistance as and when required.

PROCESSING OF THIN FILMS

In the early days of film circuit technology thin film circuits were built up by selective deposition through in-contact or out-of-contact masks. Today, however, processing is almost entirely by subtractive methods. Starting with the substrate entirely covered with the appropriate film structure those unwanted parts are removed by selective chemical etching until the required resistor and conductor pattern remains. Using the high quality photoresists developed under the impetus of the semiconductor industry, a very high definition is obtainable. The subtractive processes also allows the use of uncommitted films deposited by a specialist manufacturer.

Mask Preparation

The complexity, small finished size and the required accuracy make computer generation of the original artwork virtually essential. Because of the comparatively low sheet resistivity of thin films by comparison with the various thick film resistor pastes a high ratio of length to width is required; the resistor is usually in the form of a closely meandered pattern with many bars. Accurate hand draughting of such patterns is tedious, and even the generation of a drive tape to input to a numerically controlled draughting machine using a digitiser or by keyboard entries is a time-consuming process.

Software at various levels of sophistication is available to tackle the entire circuit layout, for example GAELIC (Graphical Aided Engineering Layout of Integrated Circuits) (2). A good compromise is to enter the co-ordinates defining the conductor pattern by digitising from a reasonably accurate hand drawing, and to add the co-ordinates of the resistor pattern (which may be 50-100 times more numerous) from a computer programme.

Thin Film Processing

Consider the meander resistor shown in figure 5.3. The resistance is:

$$R = \frac{d}{W} R_o$$

where
- d: is the total length along the midline of the resistor track,
- W: is the track width, also the track separation (in same units as d),
- R_o: is the sheet resistivity (in ohms/square)

The distance between points a_2 and a_3, a_4 and a_5 etc., is 2 W. The distance between a_3 and a_4 etc., is (B - W). Thus, the total midline length of each L-shaped section of the pattern, e.g. from a_3 to a_5, is (B + W), and hence

$$d = N (B + W) \text{ when N is the number of such sections.}$$

If N is large the slight complication due to the relative starting and finishing corners of the meander may be neglected, in which case

$$N = \frac{A}{2W}$$

thus $d = \frac{A}{2W} (B + W)$

and $R = \frac{A(B + W)}{2W^2} R_o$

The current flow round a corner is not parallel to the track boundaries, the flow lines are more closely spaced on the inside of the corner. The effect is a reduction in resistance and it can be calculated that for a right angled corner the resistance is reduced by an amount equivalent to a shortening of mid-line length by about 0.5 W. (In strict terms the equivalent midline length correction depends on the proximity of the corners. For a single isolated corner the correction is less than 0.5 W, for two immediately adjacent corners i.e. a 180° "hairpin", it is greater than 0.5 W. For two corners spaced 2 W apart the figure of 0.5 W is very accurate). There are in all 2N corners so that the equivalent length of track

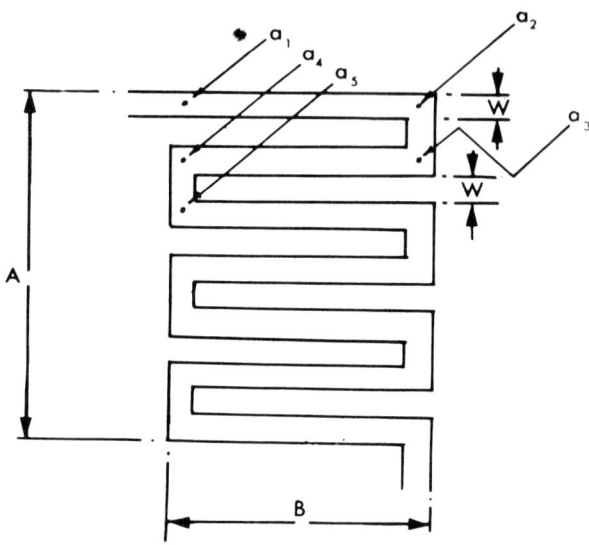

Figure 5.3 Calculation of the resistance of a meandered thin film resistor.

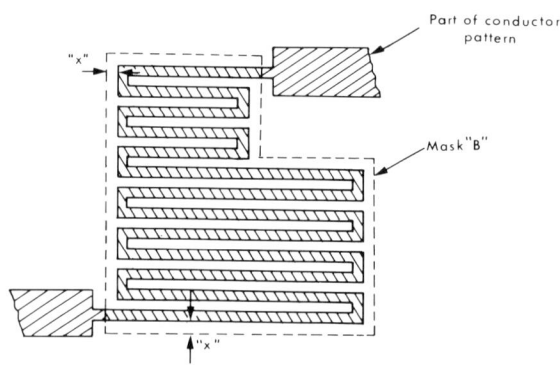

Figure 5.4 Illustration of the use of differentiating masks; mis-registration by an amount x in either direction will not affect the resistor value.

$$d' = d - NW$$
$$= d - \frac{A}{2}$$

A simpler (and more accurate) expression for the resistance is obtained from this modified length

$$R = \frac{d'}{W} R_o$$

$$= \frac{\frac{A}{2W}(B+W) - \frac{A}{2}}{W} R_o \quad \ldots \ldots (1)$$

i.e. $R = \frac{AB}{2W^2} R_o \quad \ldots \ldots (2)$

Thus the resistance of the meandered resistor having equal track width and spacing is determined immediately from the track width, the sheet resistivity and the area which bounds the resistor.

A programme GADOR (Graphical Aid to the Design Of Resistors) (3) will compute the best value of track width to fill the available area (any rectangular polygon) given the co-ordinates of the boundaries and the location of the start and finish points. It will then calculate the co-ordinates of the pattern (4N of them) and will generate a drive tape to control an NC draughting machine. The programme can be arranged to generate extra-width portions of track of appropriate length to allow for subsequent trimming.

The master artwork is best prepared by substituting a cutting knife for the draughting pen and using the same rubylith material that the thick film process employs. A scale of 20 x finished size is suitable. By changing the datum point and re-running the drive tapes multiple copies of the circuit pattern may be generated on the same sheet of rubylith, to simplify (or even eliminate the requirement for) the step and repeat process during photo-reduction. (4) Two master rubyliths need to be made to provide the two photomasks which will eventually be required.

Photoreduction and Mask Production

The backlighting frame which carries the rubylith masters and the camera must have very stable mountings and be protected from vibration. Movements of 1 micron (at final image size) are significant. An excellent lens of high resolution and flat, distortion-free field is needed and photographic emulsions of "semiconductor grade" quality should be used. The artwork must be accurately parallel to the photographic plate to avoid trapezium distortion and positioned to intersect accurately the axis of the camera.

The step-and-repeat is used as appropriate to produce a finished mask which fills, as nearly as possible the 50.8 mm x 50.8 mm film coated substrate. For a series of thin film hybrids which require more than one rectangular film circuit for their complete build-up, a separate mask set should be made for each. It is uneconomical to mix different film circuits on the same mask set.

Photoreduction and subsequent processing of the photographic plate are best carried out in high grade clean room conditions. The "standard" (silver halide) photographic emulsion is soft and easily damaged. Masks made from such plates have a finite life and several copies of each should be made. Recent developments have made available high quality glass plates with a deposited chromium film instead of photographic emulsion. These produce masks which have high contrast, are resistant to damage, and may be readily cleaned.

Resist Coating

Photoresists are organic materials which polymerise under the action of short wavelength light (blue or ultraviolet). They are most usually applied to the surface of the uncommitted substrate in liquid form; surplus is removed centrifugally by spinning the substrate plates at high speed while the volatile solvents are driven off by heating the plates to a modest temperature using infra-red lamps.

The coating process is very critical. It is carried out in a "yellow room" (familiar to semiconductor manufacturers), lit by gold fluorescent lamps which do not activate the resist. Any windows must be covered with suitable (yellow) filters to keep the lighting conditions photographically "safe".

The particle count must be kept low (Class A, 10,000), scrupulous precleaning of the surfaces to be coated is necessary, and any solids suspended in the resist solution removed by dispensing through a "Millipore" filter. Suitable filter holders for direct attachment to "Luer-Lok" syringes are available.

The photographic sensitivity of the dried resist coating (equivalent to the "speed" of a conventional emulsion) depends on its thickness. "Latitude" is limited so that careful control of resist thickness becomes a requirement. Spin speed, acceleration time, spin duration, resist viscosity and substrate temperature prior and during coating must be accurately controlled. A 50.8 mm x 50.8 mm plate, spun at 3,000 revolutions/minute for 30 seconds under infra-red heating will produce a finished (dried) resist coating about 1.2 microns in thickness.

Resist Development and Film Etching

By bringing the resist coated plate into direct contact with the mask and exposing to illumination from (say) a mercury vapour quartz lamp, selected portions of the resist are polymerised. Treatment with suitable solvents (called developers) will then dissolve the unexposed parts of the resist film, exposing the underlying metal films to later action by appropriate chemical etchants. The methods derive in large part from those used in semiconductor device manufacture. Although there may be variations in the masking sequence adopted, a suitable and convenient arrangement is as follows :

Using mask "A" (the total circuit pattern mask), all films are removed down to the bare substrate, retaining the total circuit geometry (conductors and resistors) delineated in the gold and underlying films.

Subsequently, using mask "B", (the differentiating mask), overlying films are removed leaving only the required resistive film at appropriate areas.

One advantage of organising the processes in this way is that the mask "B" need only define the overall boundary limits of the resistor. If the pattern is slightly oversize (see figure 5.4), slight mispositioning will (with appropriate layout) have no effect so that the requirement for precise registration of the two masks is avoided.

It can be seen from figure 5.4 that a slight displacement of mask "B" either horizontally or vertically by a distance less than "x" will not affect the total resistance value.

A typical process sequence is as follows :

	1.	Clean substrate
	2.	Photoresist coat
Mask "A"	3.	Expose photoresist
	4.	Develop photoresist
	5.	Inspect. Rework, if necessary, from step 21 then 1 et seq.
	6.	Etch 1 (remove gold)
	7.	Etch 2 (remove palladium)
	8.	Etch 3 (remove titanium)
	9.	Etch 4 (remove nichrome)
	10.	Inspect. Rework, if necessary, from steps 6, 7, 8 or 9 as inspection indicates
	11.	Remove photoresist
	12.	Clean
	13.	Photoresist coat
Mask "B"	14.	Expose photoresist
	15.	Develop photoresist
	16.	Inspect. Rework, if necessary, from step 21 then 12 et seq.
	17.	Etch 1 (remove gold)
	18.	Etch 2 (remove palladium)
	19.	Etch 3 (remove titanium)
	20.	Inspect. Rework, if necessary, from steps 17, 18, 19 as inspection indicates
	21.	Remove photoresist
	22.	Inspect
	23.	Clean

Subsequent Processing

Processed substrates have to be stabilised by baking in air at an accurate and uniform temperature (about $350°C$, the precise value being determined by life-test experiments) for a prescribed interval (several hours). The film resistivity increases by about 15% during this process but thereafter the effect of exposure to high temperature is substantially reduced.

Close tolerance resistors can then be trimmed using either a pulsed ruby laser operating at 1 - 2 kHz repetition frequency, or spark discharge machining. "L-cuts" or "ladder" trims can be used in a manner similar to that encountered in thick film resistor trimming methods.

Ladder trimming, where the current flow at the cut edge is zero, is to be preferred as there is evidence that the material adjacent to the cut line (kerf) is less stable than the unmachined film. Current flow parallel to long L-cuts should, therefore, be avoided.

Care should be taken to ensure, by proper layout design, that the width of the current-carrying film is not overly reduced by trimming. The current density in thin film resistors is always very high, and over-rating may result in current densities sufficient for the onset of electro-migration effects.

Out-of-tolerance or otherwise faulty units are identified (by inking or laser marking) after which the individual film circuits are separated. Scribing and breaking, using diamond scribers developed for silicon wafer division, may be used but wastage is high. The very low alkali glass does not break reliably along the scribe line. Superior results are obtained with diamond saw cutting, using a 35 micron saw running at high speed under lubricant.

HYBRID CIRCUIT ASSEMBLY AND TESTING

These processes have so much in common with thick film hybrid technology that few points need be made.

Hermetic packaging is required if the full potential of thin film circuits is to be realised. The film circuits, the active devices (usually mounted on carriers and pre-tested), monolithic capacitors and any other specialised devices are mounted individually to the inside of the package floor, using liquid epoxy-loaded glass fibre film. Interconnections, circuit crossovers, and pin connections are made, as previously stated, by microbonding using gold or aluminium wires.

The final sequence of assembly processes will, typically, be

24. Inspect. Rework as necessary
25. Electrical test
26. Centrifuge and vibration
27. Electrical test
28. Inspect. Rework as necessary (only as permitted in capability manual)
29. Final clean
30. Package seal
31. Leak test
32. Electrical test to full specification

PERFORMANCE AND APPLICATION OF THIN FILM CIRCUITS

The outstanding feature of the thin-film circuit is the stability, low temperature coefficient and accuracy of the resistors it contains. Electronic design continues to move more and more towards digital design methods for processing and the need for accurate component values in the "central" stages of an electronic circuit has largely disappeared. Nevertheless most sensors and other input devices provide an analogue output and many displays are also, preferably, analogue in their presentation. At the input and output of electronic equipments accurate A-to-D and D-to-A conversion is necessary. These converters are well known, but several similar requirements arise, such as synchro angle encoding in Cartesian form (an electronic equivalent to the Scott transformer) and generation of accurate non-linear functions (5).

As an indication of the required accuracy and stability, table 5.1 gives the resistor specification of A-D converters of various bit numbers. Although the film resistivity is low, adequately high resistance values may be generated without allocating unduly large areas of substrate. Reference to equation (2) shows that, for

TABLE 5.1

Thin Film Resistor Specification for A-D Converter Applications

Number of bits	8	10	12*	14
Required accuracy (MSB Resistors) (%)	0.4	0.1	0.025	0.006
Resistivity (Ω/sq)	50–500			
Temperature coefficient of resistivity TCR (ppm $^\circ C^{-1}$)	100	70	50	35
Operating temperature range ($^\circ$C)	−65 to +125			
TCR tracking (ppm $^\circ C^{-1}$)	20	5	1.25	0.3
Power density (W cm^{-2})	25			
Stability, absolute (% (1000h)$^{-1}$)	0.02			
Stability, ratio (% (1000h)$^{-1}$)	0.001			

* If the number of bits is greater than 12, the converter accuracy begins to be limited by errors in the switches.

20 micron track width and spacing, 300 ohm/square resistivity, a resistance of 37 megohms can be accommodated in each square centimetre of substrate. At the low resistance end, resistors become bulky and approximately 3 ohms is the minimum practical value.

Temperature coefficients are better than 10 ppm over the range -40°C to +125°C. Tracking of resistors on the same substrate is about three times better than this. In the mid-range (100 ohms - 10 megohm) trim accuracy is 0.05%. Stability at maximum rating (i.e. with voltage applied and film temperature 125°C) is better than 0.15% for 1000 hours (i.e. $\Delta R/R < 0.0015$ after 1000 hours). Stability of resistor <u>ratios</u> is, with carefully chosen layout, five times better than this.

Thus it can be seen that present day thin film circuits and networks can comfortably meet the requirements of A-D converters of 10 bit accuracy while 12 bit accuracy can be obtained if the upper limit of operating temperature is restricted and some decreased yield due to trimming errors (in the more significant resistors) is acceptable. The technology and performance continues to improve and unconditional 12 bit accuracy, at high temperature and over long service life is a reasonable objective. At this point other problems, such as variations in the forward resistance of the switching devices, begin to limit the overall performance.

Though thin film circuits are expensive in relation to their thick film counterparts the costs are increasingly offset by improvements in the photolithography and better process yields. Improved resolution and consequent reduction in (say) resistor track width increases the number of circuits derived from a single plate and thus lowers the unit cost.

A promising direction of development is the "hybrid hybrid", an integrated circuit in which the majority of the conductor pattern and the less critical resistors are realised in thick film technology and thin film circuits, which generally need be only a few square millimetres in size, are used in the manner of other add-on components to provide the critical resistors where precision, stability and low temperature coefficient of resistance are required. This approach deploys fully the particular advantages of the thin film technology without greatly increasing the production cost of the complete circuit.

6
The Design of Thick Film Hybrid Circuits

I. D. SALISBURY

INTRODUCTION

The hybrid manufacturer offers service of interconnecting and packaging of components which are available on the open market and by special techniques assembles these components onto alumina substrates with interconnection formed by thick film conductors, solder, resin and, in some cases, fine wires. Generally, but not always, thick film resistors are also used. The final assembly is then encapsulated and tested to meet the agreed electrical and mechanical specification to a defined environmental performance. The task of the hybrid designer and manufacturer is to provide this service to the customer at a commercially acceptable price and mutually agreeable timescale.

The starting point for the designer is whether to use hybrid technology at all. This may seem a strange comment but in the past many circuits have been realised in hybrid technology just for the sake of using this technology, when a printed circuit board realisation, a custom semi-conductor device or some other technology would have been correct choice. It is therefore necessary to define in detail the objective for the hybrid design with the purchaser since without identifying the objective, the product may well not meet the requirements which are going to be placed upon it. Although this might be considered to be stating the obvious, it is important to recognise that the flexibility of the thick film process can lead to problems of definition and that

Designing Hybrid Circuits

these areas of potential ambiguity must be eliminated at the outset.

Some of the common reasons for deciding to use hybrids are :-

> Size reduction
> Cost reduction
> Circuit performance
> Thermal matching
> Increased reliability
> Simpler overall system and circuit design

These are based on requirements of the circuit or system being designed. There are other commercial reasons why hybrids should be used, such as

> Low capital investment
> Easier to update and change technology
> Ability to purchase fully tested unit
> Reduced goods inward inspection
> Fewer supplier to control
> Lower purchasing cost
> Fashion

Having established that there are sound reasons for using the technology, the next stage is to find the equipment practice to be employed. This will cover such areas as

> Temperature range experienced by the hybrid
> Package, shape and size
> Identifying pin out position and marking
> Package style (e.g. single in line resin coated)
> Thermal design
> Electrical screening
> Environmental levels
> Electrical performance specifications
> Methods of mounting

From these boundary conditions the hybrid designer can now select the technology to suit the requirements.

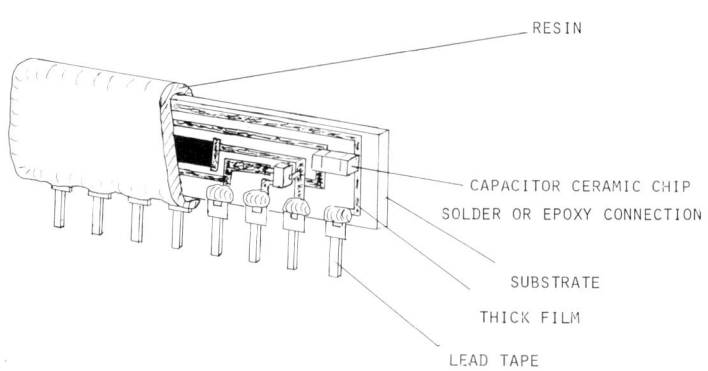

Figure 6.1 A single-in-line resin coated hybrid structure.

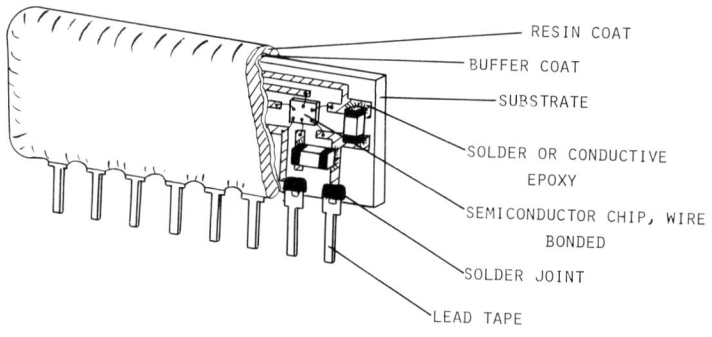

Figure 6.2 A single-in-line resin coated assembly using unencapsulated semiconductor devices.

Designing Hybrid Circuits 85

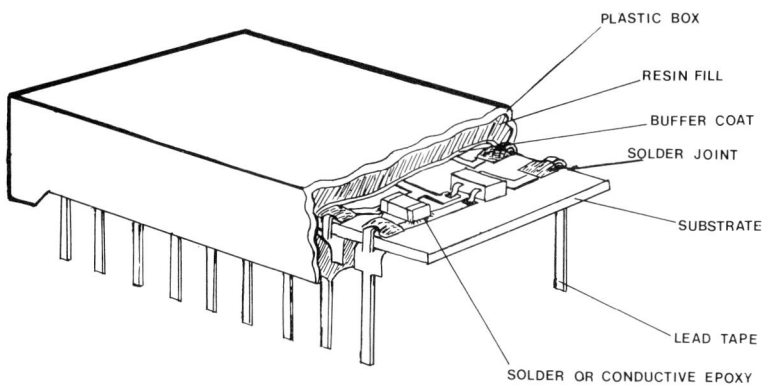

Figure 6.3   A dual-in-line resin coated package.

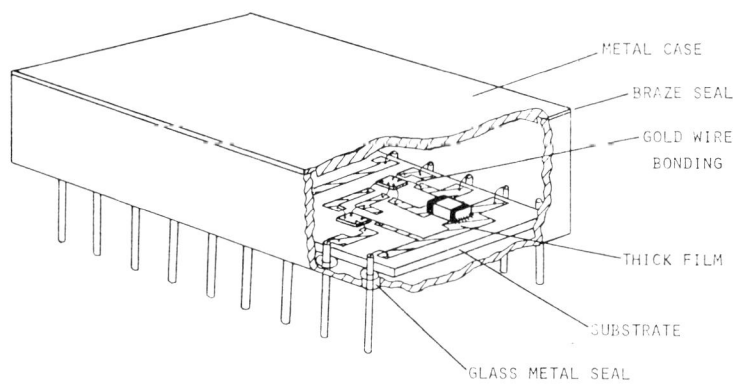

Figure 6.4   A welded hermetic high-reliability metal package.

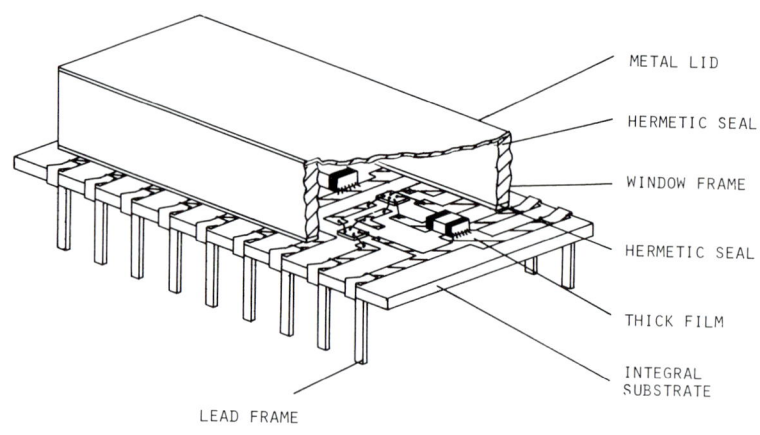

Figure 6.5 Typical integral substrate package.

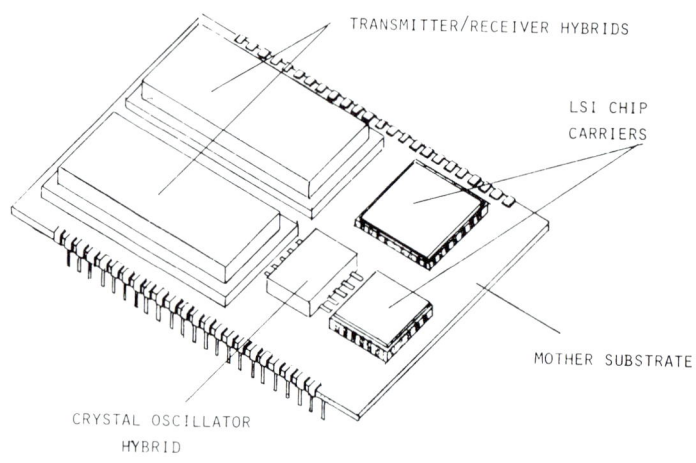

Figure 6.6 A complex hybrid assembly using a variety of package concepts.

CHOICE OF PACKAGE

One of the most important aspects of the thick film design is the choice of package. The different methods of packaging and the mechanics of their assembly are discussed in chapter 7. Some of these package styles will now be reviewed and the way in which they impact on the design of the overall hybrid presented.

Figure 6.1 shows a typical single in line resin coated structure which is generally satisfactory from $-40°C$ to $+100°C$. With care in the design and the choice of correct materials, this can be extended to an operational temperature of $125°C$. Figure 6.2 depicts a single in line structure using a resin coated chip and wire technique. This is similar to the previous package but a higher component density can be achieved. To reduce the problems of wire bond failure due to the mechanical stress imparted by the hard resin, a flexible buffer coating has been placed between the fragile wire bonds and the resin encapsulation. Figure 6.3 shows a dual in line, resin coated hybrid. It is very difficult to resin encapsulate the dual in line style by straightforward dipping procedures and so a plastic box has been placed over the circuit and used as a mould for the resin.

The high quality military products are in general encapsulated in metal enclosures with glass to metal seals and a typical package is shown in figure 6.4. After the assembly of the components, the package is sealed by either welding or brazing. The metal package is a high cost item and the packaging will be a significant part of the selling price for large hybrids.

The package style known as Integral Substrate Package (ISP) as illustrated in figure 6.5, uses the thick film conductors and dielectric to form a lead out from the hermetic enclosure and hence replaces the need for glass to metal seal and a metal package. It also has the advantage of not having a resin seal between the substrate and package which can significantly improve the thermal performance of a package.

In figure 6.6 a large area hybrid approximately 3" x 2" is presented that combines the use of chip carriers and ISP hybrid packages, both of which are surface mounted onto a multilayer

thick film substrate. The advantage of this technique is that the
active components in the chip carriers can be tested prior to
mounting on a large area mother board and high yields at manufacturing stage can be obtained.

   To summarise, as a general rule resin encapsulation will cover
-40°C to +100°C and metal case -65°C to +150°C. It must be
remembered however that the temperature range can be reduced if
the individual components and materials used in the hybrid assembly
are unable to withstand these temperature extremes.

   It is obvious therefore that cost, environment and application
are related in a complex manner and in figure 6.7 an attempt
has been made to demonstrate some aspects of this relationship. It
can be seen that if cost is used as a base line then the semiconductor, chip and wire bonding in the metal package is the
highest cost and solder assembly in plastic encapsulation is at the
lower end. The term grade refers to the quality level but as the
technology matures it is becoming much more difficult to relate the
boundaries of package technology to quality levels since resin
assemblies are now being accepted into military and telecommunication equipment where previously only a fully hermetic device
would have been allowed.

CHOICE OF COMPONENTS

   Once the style of package has been decided, it is necessary
to choose the electrical components to be used in the circuit. The
selection of the right type of component from the vast range
available has to be made with great care if a cost effective
module is to become a production item. The low cost components
which are now available are, in the main, those which are made
in large quantities to suit the conventional printed circuit board
technology and historically this has meant they will have wire
terminations designed to be pushed through holes in the PCB and
soldered on the opposite side of the board to the component. As
the manufacture of holes in ceramic substrates is very expensive
only a small selection of this wide range are suitable for use in
hybrids. Because of their low cost, some wire ended components
have to be used but these should have the wire formed so that they
can be surface soldered. With the more recent trend in using

Designing Hybrid Circuits

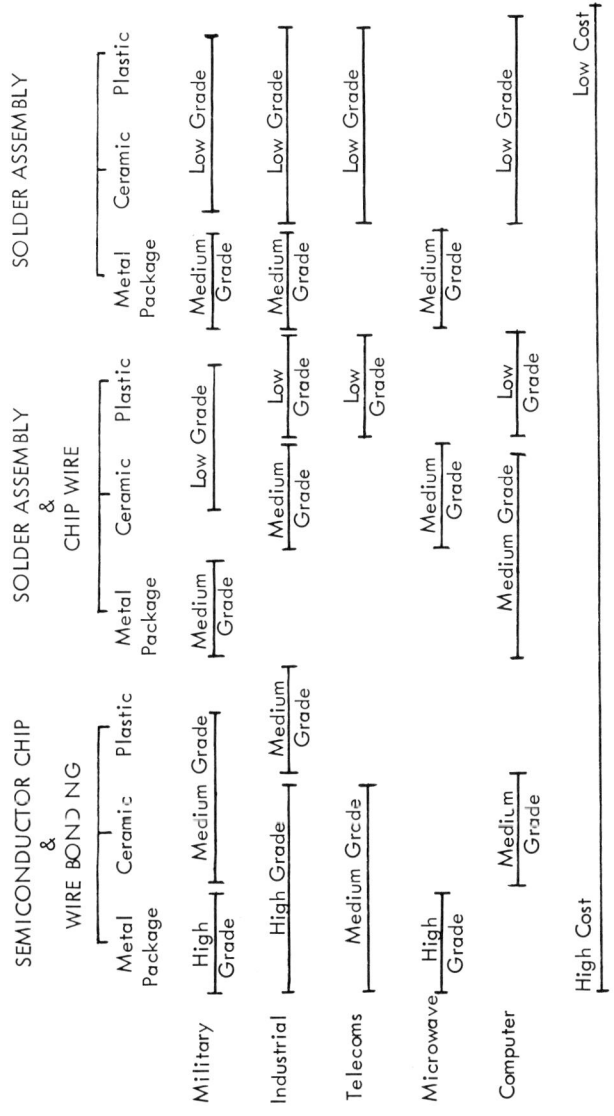

Figure 6.7 A guide to the application areas of the various package concepts.

surface mounted components generally, a more extensive range of new components is becoming available and as their use increases so the cost will fall. Some typical package outlines are shown in figure 6.8. Figure 6.9 gives some predictions made for the relative prices of chip carriers, ceramic chip capacitors and semi-conductor devices in SOT packages. If these predictions are correct, then circuits designed in 1983 employing these devices would expect to enter production in 1985 using what would then be cost effective components.

The importance of choosing the cost effective components can be seen in the costing of a typical active filter such as the one shown in figure 6.10. This circuit uses 19 resistors, 6 capacitors and 4 integrated circuits. A costing has been carried out using manual assembly and also automatic assembly. The area of cost reduction that can be achieved by using devices which are designed for automatic assembly can be gauged from figure 6.12. This breakdown of costs shows the importance of selecting the right components for the circuit since the selection greatly affects the ongoing cost and reduces the risk of redesign being required in order to maintain the competitive price of the product.

Now that the physical size of the hybrid and type of components that are going to be used have been chosen the layout of the circuit design can commence. If the circuit has been designed by the customer using computer aided techniques, then analysis of the effect on circuit performance such as conductor resistance, capacitance coupling between adjacent conductors, capacitance of crossovers, and capacitance to the ground layer should not cause a problem. Looking at the various components which can form the hybrid, some significant parameters are worth reviewing.

THICK FILM CONDUCTORS

Thick film conductors are available in three basic materials : gold, silver and copper. Gold and silver can be blended with platinum and palladium for various applications, but this does have a significant effect on the resistance as can be seen from figure 6.11. An example of this effect would be that the resistance of a 2" long conductor of width .01" could lie between

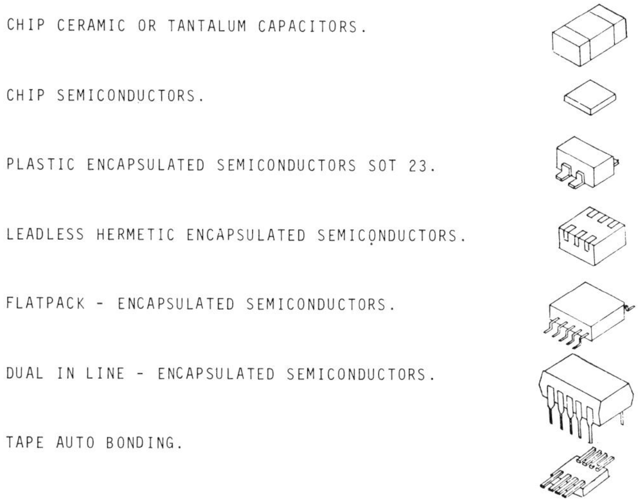

CHIP CERAMIC OR TANTALUM CAPACITORS.
CHIP SEMICONDUCTORS.
PLASTIC ENCAPSULATED SEMICONDUCTORS SOT 23.
LEADLESS HERMETIC ENCAPSULATED SEMICONDUCTORS.
FLATPACK - ENCAPSULATED SEMICONDUCTORS.
DUAL IN LINE - ENCAPSULATED SEMICONDUCTORS.
TAPE AUTO BONDING.

Figure 6.8 Schematic diagram of some of the component styles available.

CONVENTIONAL HERMETIC PACKAGE
2"x2"AL2O3
SOT 23
CHIP CAPACITOR
TAB
SO 14
HERMETIC CHIP CARRIER
COST
10
9
8
7
6
5
4
3
2
1
1979
1980
1981
1982
1983
1984
1985
YEAR

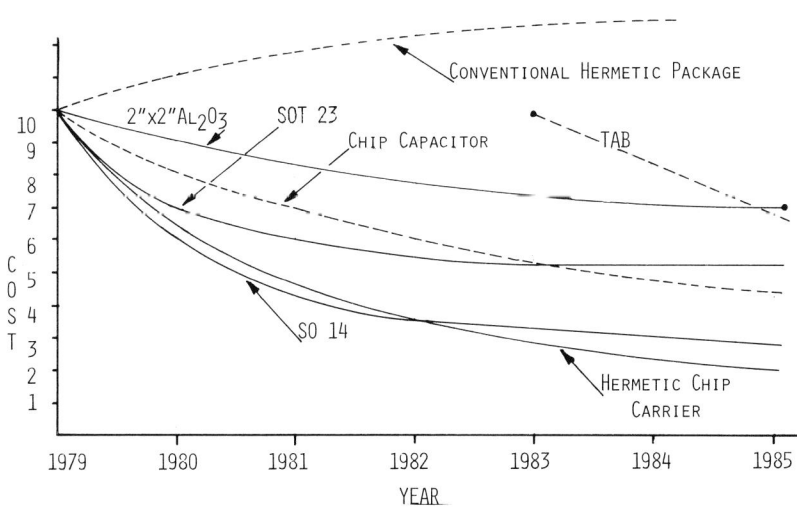

Figure 6.9 Projected costs for various hybrid components.

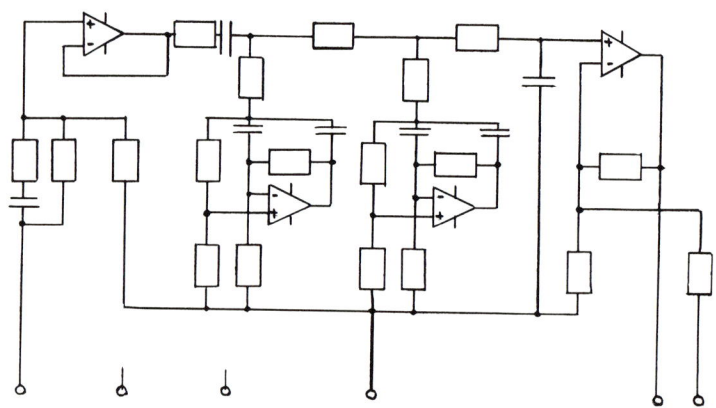

Figure 6.10 A typical circuit suitable for hybrid manufacture.

CONDUCTOR MATERIAL	RESISTANCE milliohms/square
Gold	2.5 to 6
Gold Platinum	15 to 110
Gold Palladium	13 to 80
Silver	1.5 to 3
Silver Platinum	3 to 15
Silver Palladium	15 to 40
Copper	1.8 to 4

Figure 6.11 Typical resistance range of various thick film conductors.

Designing Hybrid Circuits

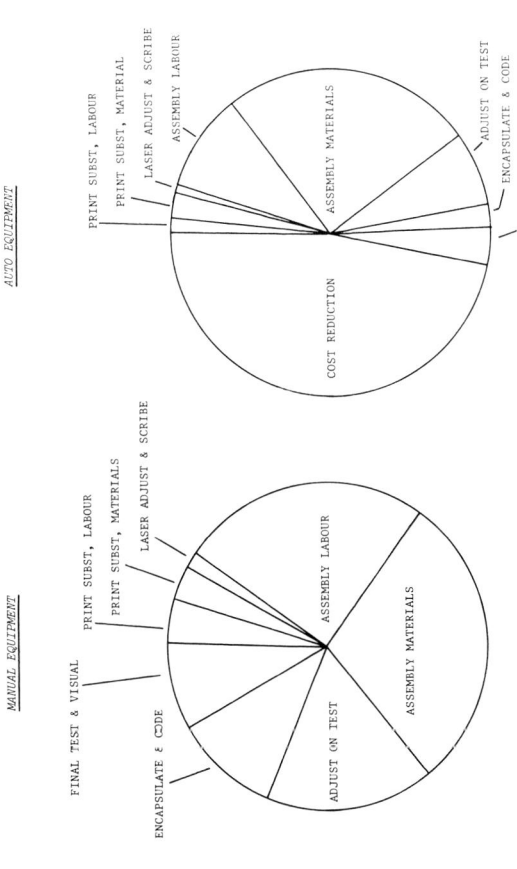

Figure 6.12 An estimate of the cost reduction that can be achieved in the manufacture of the circuit shown in figure 6.10 using a cost-effective choice of components and assembly methods.

3 and 20 ohms. This can cause a number of problems such as volatge modulation on power rails, reduced speed of operation of high speed logic circuits and incorrect voltage ratios etc.. As the temperature coefficients of conductors can be as high as 3000 ppm/°C, errors in the voltage ratios in critical areas as a function of temperature can also occur.

For use in solder assembled hybrids, silver-bearing type conductors are generally employed either as a binary alloy such as silver palladium or, if better adhesion and solder leaching resistance is required, then as a ternary silver palladium platinum alloy. There is a wide range of these materials available on the market, but care should be taken in understanding the process sequence. If the conductors experience a large number of refirings, loss of adhesion and reduced solderability can occur.

WIRE BONDING

Wire bonding is common in two forms : ultrasonic aluminium and thermosonic gold ball bonding. In general gold-bearing conductors are used for both although recently success has been achieved with bonding onto silver-bearing conductors. With both these techniques yield problems on bonding can occur if the design of conductor is incorrect. Figure 6.13 shows a typical conductor pattern for bonding a semi-conductor chip. Higher bonding yields will be obtained by wires entering the thick film conductor from the end than will be achieved by the wire bonding across the conductor. When bonding onto a narrow conductor (0.15 mm wide, say) which, due to surface tension, will not be level, there is a chance that the bonding tool will not be in contact with the bonding wire if the tool hits the conductor first. Therefore a good bond will not be formed since the energy is not dissipated between the wire and the thick film conductor. The position of the bonding tool can be seen to be critical and the track pattern should be designed to optimise the yield on bonding.

THICK FILM CROSSOVERS

Crossovers formed by thick film techniques have a capacitance of about 2 pF per square millimetre. For military applications, crossovers are generally made using gold conductors, while in

Designing Hybrid Circuits

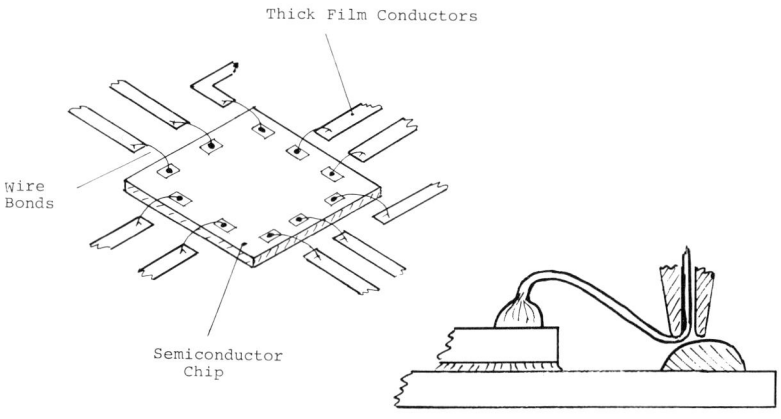

Figure 6.13　Typical conductor layout necessary to achieve a high yield on wire bonding.

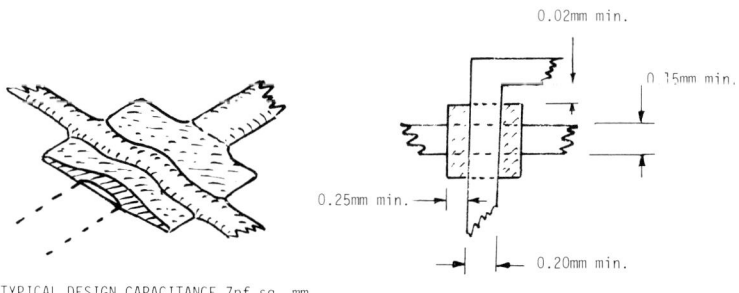

TYPICAL DESIGN CAPACITANCE 7pf sq. mm.

Figure 6.14　Typical crossover design.

industrial type products, silver bearing conductors are used, but these have been known to give rise to leakage under conditions of high humidity due to silver migration. Recent studies with carefully chosen dielectric materials have shown that silver-bearing conductors can be used in military applications and give a satisfactory performance but care must be exercised in choosing the right dielectric and correct composition of conductors to achieve a reliable crossover. A satisfactory design of a crossover is shown in figure 6.14.

RESISTORS

Typical thick film resistor performance is shown in figure 6.15. Better stability, tolerance and TCR can be obtained by careful selection of materials, resistor design and adjustment technique. New inks are becoming available which claim better stability and TCR. If these boundary conditions are used in the design, high yields and good long term performance circuits can be expected.

Resistor design and adjustment techniques are important factors in obtaining optimum stability. In general, when a thick film resistor is produced, due to a wide range of effects in both the printing and firing procedures and production spreads of resistor inks themselves, a 50% variation in the as fired value can be expected and must be allowed for. This means therefore that if a 1% tolerance resistor is required, accommodation has to be made for a 50% adjustment. The adjustment is achieved either by laser machining or air abrasion to remove resistor material and hence alter the current flow. This in turn changes the resistance as measured from the terminals. A typical distribution curve for thick film resistors is shown in figure 6.16, and some typical trimcuts in figure 6.17.

The detailed design of the adjusted resistor must take into account voltage stress, power stress, stability with life etc.. These are critical to the method of adjustment and as a general rule the tighter the tolerance and the higher the stability required, the slower rate of change of adjustment the better. Also large resistors give better yield and stability. Figure 6.17 shows the various methods by which the adjustment cuts can be made and the percentage change of resistor with depth of cut. Where possible,

Resistivity	10Ω/□ to 1MΩ/□ In Decade Steps
Adjustment by Laser	Any Value up to 30MΩ
Tolerance - Absolute	±0.5%
- Matching	±0.5%
Stability	10K hours
70°C Film Temperature	<100Ω/2.5%
	>100Ω/0.8%
125°C Film Temperature	<100Ω/3.0%
	>100Ω/1.0%
TCR - Absolute	±100ppm/°C −55 to +25°C/+25 to 125°C
- Tracking	±50ppm/°C Same Ink

Figure 6.15 Performance limits of thick film resistors.

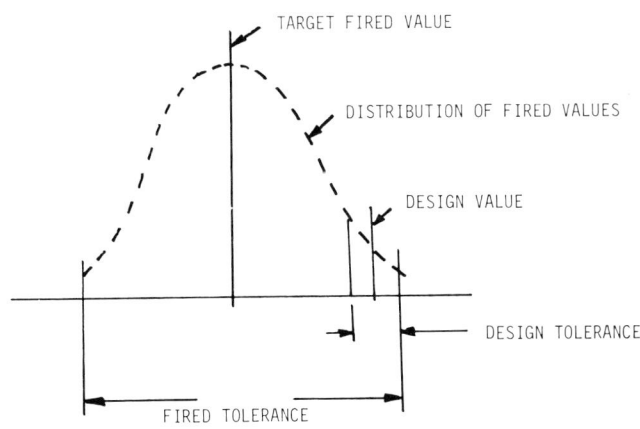

Figure 6.16 Distribution curve of as-fired resistors.

98 I.D. Salisbury

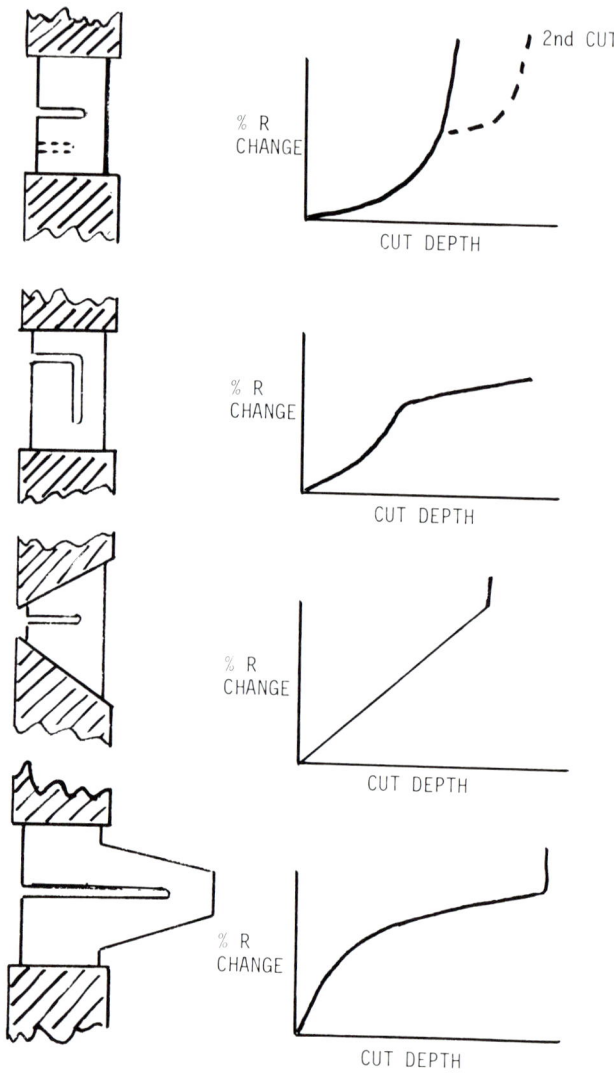

Figure 6.17 Some frequently used trim-cuts.

all resistors designed to obtain the best stability and lowest final tolerance need to be adjusted very slowly. Care must be taken to ensure that resistors remain within the paste manufacturer's specified temperature. Local "hot spots" caused by current crowding after trimming can cause the resistance to drift.

THERMAL DESIGN

The reliability of a hybrid and the device operating temperatures are closely related. As an example, a ceramic capacitor operating at maximum volatge stress and 25°C, would have a failure rate of 55 devices per 10^9 component hours, at 100°C the failure rate would be 340 devices per 10^9 component hours, and at 125°C, 1700 devices per 10^9 component hours. Thus a factor of 6 worsening in reliability between 25 and 100°C and a factor of 31 between 25 and 125°C must be expected. Similar trends can be monitored for other components and therefore thermal design is a very important factor in obtaining high reliability hybrid circuits. There is also an important effect caused by the mechanically induced stress due to the wide temperature variations and differing thermal coefficients of expansion.

When using items such as SOT23 transistors and similar devices, it is important to position these components on a cool part of the substrate and, for example, when these components are dissipating power, solder coat the conductor tracks to act as heat spreaders. In this way a very significant reduction in junction temperatures can be realised in the semi-conductor device.

When semiconductor devices or substrates are glued in position, the choice of the glue material is important. Figure 6.18 shows the thermal conductivity of the various materials which might be used to mount the devices and it can be seen that there is a very significant difference between the thermal conductivity of a conductive resin and a eutectic bond. If resin is used to attach the semi-conductor chip to the substrate, the temperature of the device can be very different to that found when a metal system is used. These effects must be taken into account when calculating the junction temperature, since as well as increasing the failure rate, a high junction temperature may cause the device to operate on a different part of the characteristic to that which was

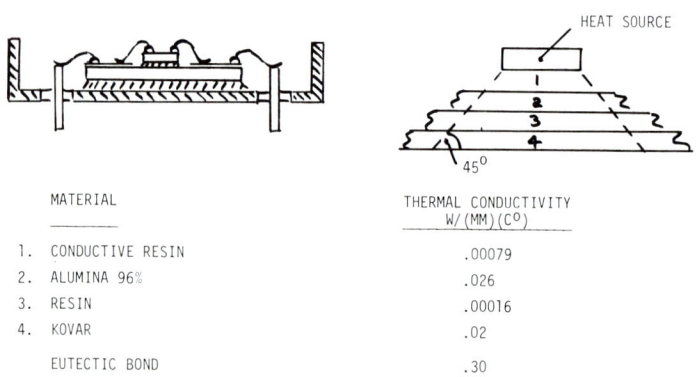

MATERIAL	THERMAL CONDUCTIVITY W/(MM)(C°)
1. CONDUCTIVE RESIN	.00079
2. ALUMINA 96%	.026
3. RESIN	.00016
4. KOVAR	.02
EUTECTIC BOND	.30

Figure 6.18   Thermal conductivities of various materials used in hybrid packages.

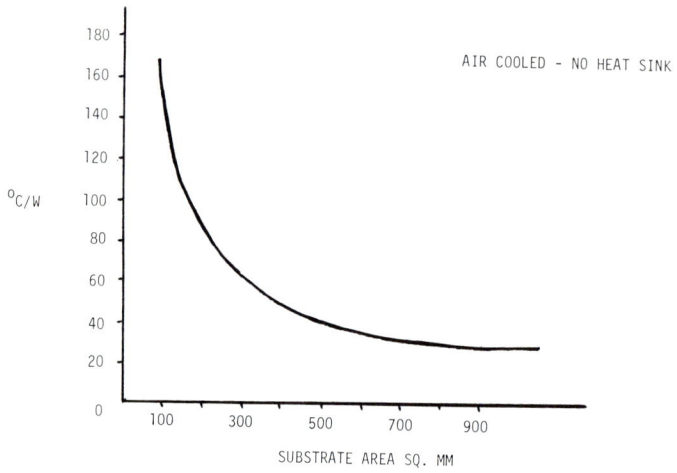

Figure 6.19   Thermal resistance of packages in free air.

proven in a bread-board design using pre-packaged components. It must also be remembered that when conductive resin is used to mount semi-conductor chips it can add up to 5 ohms resistance in the conduction path and limit the operation to 125°C. Higher conductivity resin systems are available but the choice of which system to use is a difficult one given the various operating conditions.

Typical design characteristics for the thermal resistance of packages to free air are shown in figure 6.19.

CERAMIC CAPACITORS

As discussed in chapter 4, ceramic chip capacitors are available in three main dielectric types. These are known as :-

NPO with dielectric constant 25 to 100
Stable K with dielectric constant 300 to 1800
High K with dielectric constant 2500 to 15000

The common characteristics are given in the suppliers data sheet, but a less well known characteristic which can cause problems is dielectric adsorption. This can be a significant parameter in pulse circuits and applications where rapid charge and discharge characteristics are important. In hybrid design, in general, the two dielectric materials which it is possible to obtain are ceramic or tantalum.

Capacitor ageing, particularly with ceramic devices, is important and typical ageing characteristics are shown in figure 6.20. The ageing starts when the capacitor has been heated above the Curie temperature of approximately 180°C. Thus, when the capacitor is soldered into the thick film assembly, the capacitor will be at zero time and should be left at least 100 hours before the circuit is subjected to electrical adjustment or fine electrical testing so that three decades of ageing have occured and the resultant change between 100 hours and the 10,000 hours would only be 2.4% for a stable K dielectric. The assembly of ceramic capacitors to thick film circuits is well established. The advantage of this component is that it has a similar expansion coefficient to the 96% alumina substrate and electrical connections

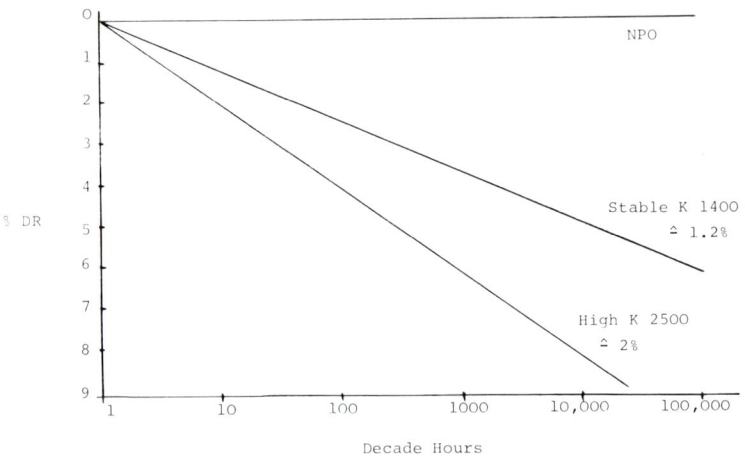

Figure 6.20   Ceramic capacitor ageing curves.

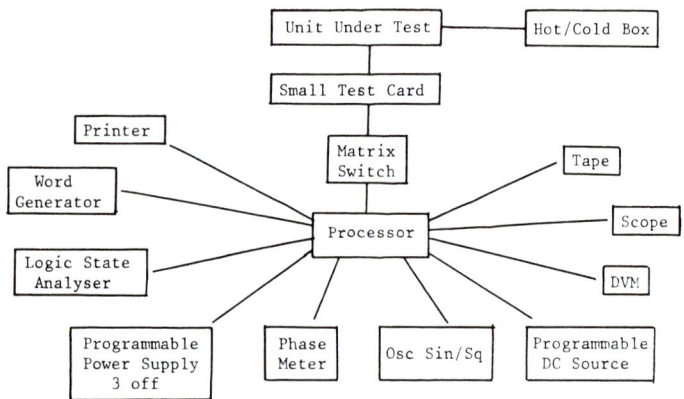

Figure 6.21   Schematic diagram of a typical automatic test station.

are made either by solder or conductive resin. Correct layout of
the hybrid can assist in the assembly by providing properly designed
solder pad areas which will self-align the chip components when
the solder is reflowed. The cost effective use of ceramic capacitors
is to design, if possible, using NPO devices below 1000 pF and a
stable K below 50,000 pF. In some cases however wire ended
components need to be used as they are frequently of lower cost and
often devices are only available in this form. They can prove
unreliable when surface mounted due to mechanical stress induced
by thermal mismatch between the substrate and the component. To
reduce this, flexibility should be left in the leads so that the stress
applied to the component itself is minimized.

HYBRID MOUNTING TO PCB

The hybrid assembly must also be designed to withstand the
thermal stress induced when it is soldered to the printed circuit
mother board. Such effects can be very significant when the thick
film package is large, since the coefficient of expansion of metal
hybrid packages and ceramic assemblies are both about 6×10^{-6}
and the printed circuit board expansion coefficient could be as high
as 24×10^{-6}. Problems and mechanical damage may occur if the
effect is not allowed for in the design.

ELECTRICAL TESTING

The component and thick film aspects have been covered so far,
but the hybrid manufacturer is offering more than interconnection
and package service to the Industry. At the same time, he has to
be capable of supplying an electrically tested design. The range
of tests required is formidable and covers resistance, frequency,
voltage, current, phase, delay, absolute gain, differential gain,
AC noise, DC offset, logic analysis, propagation delay, transfer
accuracy, droop rate, settling time, transients and many more. In
addition to these measurements, all the components used in the
assembly must also be tested. A very comprehensive range of test
equipment, which is a high capital investment, is therefore required
and in some cases little used. Once the test equipment is
purchased the next problem is the interface test box between the
hybrid and the instrument testing system. There are four main
approaches to this; (a) small custom built test box which can

connect standard test instruments and power supplies to the hybrids by means of a hand operated switch, (b) custom built test boxes which include power supplies and signal generators etc., with a hand operated switch giving a go/no-go output, (c) a simple test card which plugs into a matrix switch coupled to a processor, that in turn, calls up standard test instrumentation and (d) an automatic test equipment system (ATE). Figure 6.21 shows a schematic diagram of such a test station. There is no single system that can be called correct since the requirement depends on the range of products and the quantity levels of these products but with the recent advances in micro-processor techniques, relatively low cost automatic test stations can be built. This has the advantage that with the correct software, the computer can advise the design engineer and the quality engineer of the electrical test results with respect to the test limits and any adjustments which need to be made to ensure high yields. Typical process routes are given in figure 6.22 and the stages at which electrical tests are made are indicated. At each of these test points rejects are likely to occur, with most rejects occuring at the adjust on test and first electrical test. The more complex the circuit, the lower the yield and therefore it is necessary to be able to diagnose quickly the defective device in this complex circuit and to be able to replace it. The diagnostic approach can be made easier if the circuit is partitioned by the designer so that it can be tested in sections, either by bringing out test points on unused pins, or including removable links within the circuit to split functions. These low yields can have a very significant influence on the final pricing of the hybrid assembly and it cannot be stressed too strongly that the correct design can significantly affect the cost and hence viability of the final product.

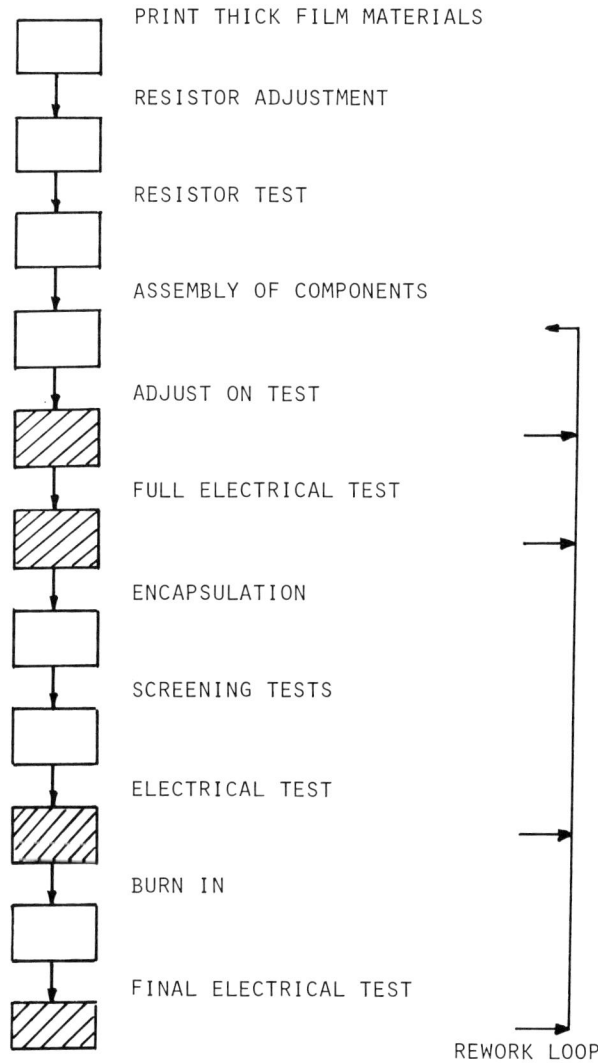

Figure 6.22 Process route for a hybrid showing the occasions on which the hybrid will be tested and also the rework loops.

7
Packaging Hybrid Circuits

B. C. WATERFIELD

INTRODUCTION

In this chapter means of protecting the completed hybrid circuit and providing a method of presenting it to the board onto which it will be mounted will be considered. The degree and resultant type of protection will depend on the **environmental** or mechanical needs of the function for which the hybrid is required. Whilst film circuits are relatively new, during their life so far there have been many changes of packaging philosophy both in terms of construction and the reasons for choosing a particular method. Because many applications have designs frozen for several years it is necessary therefore to be familiar with a considerable number of package constructions and to understand their relative merits.

SELECTION OF TYPE OF PACKAGE

The selection of the packaging for a particular circuit will depend on a number of facts and there will almost certainly be constraints placed upon the package by the customer such as overall size, pin-out arrangement, identification marking, operating conditions etc..

Figure 7.1 summarises the principal options open to the designer and should help place the various components and packaging concepts in perspective. It is essential that in the design of hybrid circuit the packaging needs must be given careful consideration and the design aspects of the package are

Packaging of Hybrid Circuits 107

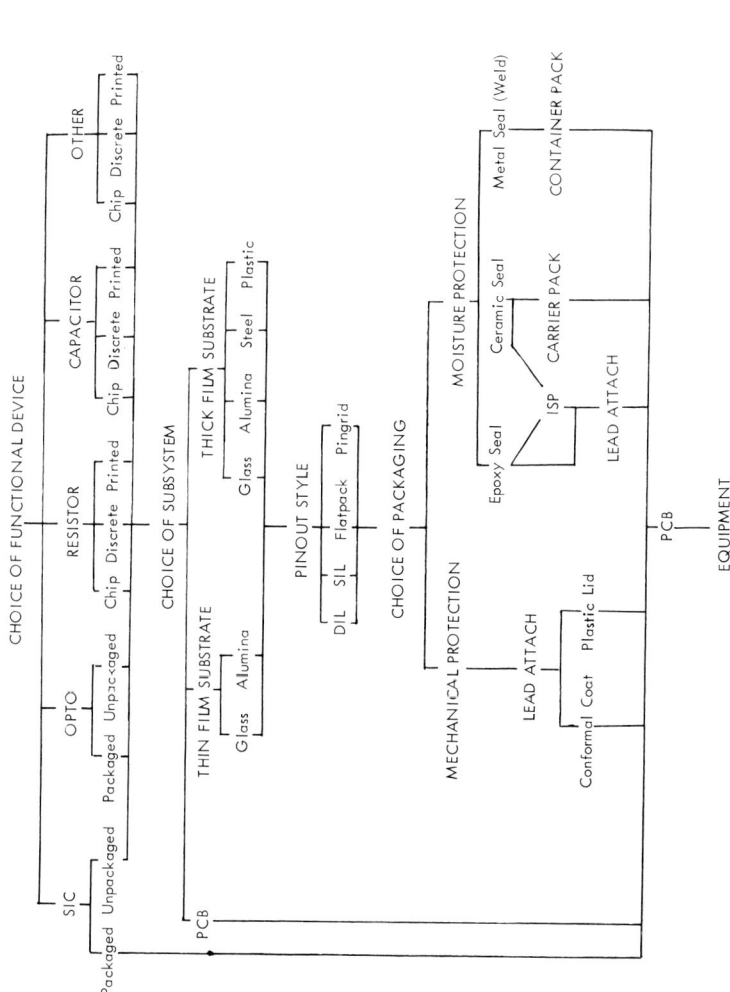

Figure 7.1 Packaging options available to the hybrid engineer.

covered in chapter six. Although many circuits can be used without any packaging at all, a large percentage will need some sort of protection, so it is first necessary to investigate the reasons for packaging. These will generally fall into one or more of the following.

(a) Protection from adverse ambient
(b) Protection from mechanical damage
(c) Aid to utilisation
(d) Displaying information

Once the need to provide some form of protection is established, the type of package can be selected.

DEVICE PACKAGING

Before considering a package for the complete hybrid, the active devices must first be examined since these are the most critical components. They can be procured in several forms.

(a) Bare chip
(b) Passivated bare chip
(c) Plastic encapsulated chip
(d) Mounted in chip carrier - i) Plastic/Fibreglass
                                                ii) Three/Two layer hermetic
                                                iii) Single layer hermetic/ moisture proof
(e) Leadless Inverted Device (LID)
(f) Flatpack
(g) Tape Automated Bonded (TAB)

Other components such as chip capacitors, inductors, resonators etc. can also influence package selection due to their physical size and weight or terminations provided.

PACKAGE CONSTRUCTION

It is possible to divide the package construction up into three broad areas :

Non-hermetic

In this case the completed circuit has lead-clips soldered to it in either a DIL, SIL, or 'butterfly' format. It is then either covered with plastic powder in a fluidised bed and heated, or alternatively dipped into molten plastic to provide protection. This form of packaging has become more far reaching in its application since the advent of devices encapsulated in chip carriers where the joints to the substrate are soldered and are not disturbed by the coating process. Chip carriers are also completely hermetic. This type of coating would not be possible with a bare chip placed directly on the substrate since the coatings are, to various degrees, non-hermetic. A number of materials are available and manufacturers are keen to claim improved moisture resistance.

The irregular surface presented by this technique may not be desirable since it makes marking difficult and does not have aesthetic appeal. These obstacles can be overcome by placing a plastic cover either outside the clipped on leads or between them. Noryl or other glass or ceramic filled nylons are used that provide rigidity, reasonable temperature resistance (180°C) and a close thermal match to the substrate as illustrated in figure 7.2.

Near Hermetic

In this case a substrate is prepared in the same manner to the non-hermetic package but a lid of similar material to the substrate - alumina - is placed over the circuit and joined to the substrate using filled epoxy preform or filled epoxy coating and then oven cured to about 150°C. Here care should be taken to ensure a thin glue-line is maintained which may require the mating face of the ceramic lid to be ground flat.

Other alternatives have also been investigated using epoxy or silicone joining materials and lids or frames made from fibre glass, metal or glass, all chosen to have a similar expansion coefficient to the substrate. Construction examples are shown in figure 7.3.

Hermetic

Before looking at the package constructions themselves it will first be necessary to examine the various methods available for achieving the final seal. It is then possible to appreciate the

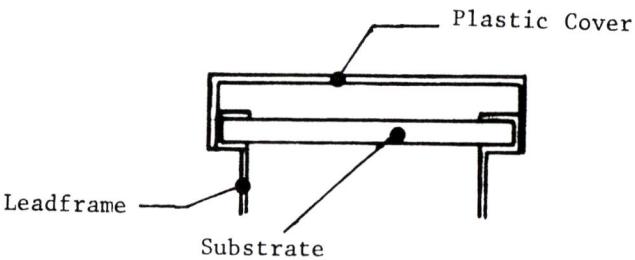

Figure 7.2 Typical plastic packaged hybrid

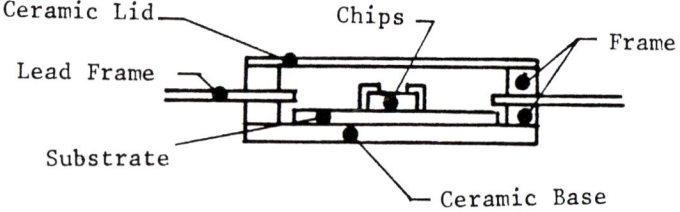

Figure 7.3 A "near" hermetic hybrid package using a glued-on ceramic lid.

SEALING METHODS FOR HERMETIC PACKAGES

To achieve complete hermeticity within a package the final seal between the base and its lid can be affected with three basic techniques :

(a) Glassing
(b) Soldering
(c) Welding

Glass, particularly with packages of ceramic construction, offers great potential in that the material is low cost and when flowed it is non porous. However, the package will generally be subjected to peak temperatures in the excess of $400^\circ C$ in a nitrogen atmosphere. This method can be used successfully on small packages such as chip carriers but it is not favoured for large, complex hybrids.

Soldering for many years has been the accepted method of sealing metal or ceramic packages but the risk of detached solder balls or unacceptable fluxes have all but excluded its application other than when gold alloys, such as AuSn, can be used since these do not require a flux and are supplied as preforms of controlled dimensions. Hand solder joints can be achieved successfully if care is taken to prevent any movement of solder into the package.

Welding of metal lids to metal packages is currently the most widely employed method of achieving reliable lid seals. Two techniques are used at present but other interesting possibilities include laser welding. This would be particularly suitable when packages of irregular shape were required.

Projection Welding is a type of resistance welding where a projection is provided either on the lid or mating face of the package. The lid must make overall contact with the base. Pressure is applied to ensure good contact, and a high current from a discharging capacitor is passed through the joint. There is some collapse of the lid at the projection point and the seal is

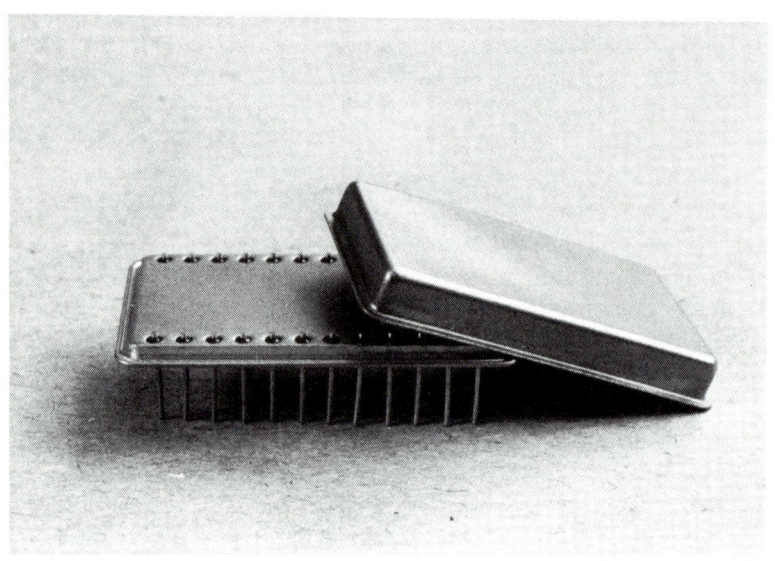

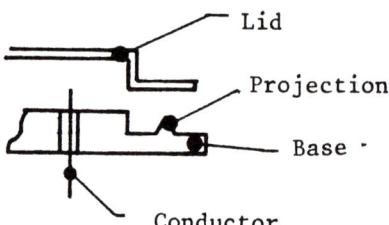

Figure 7.4 A projection welded platform package
          Above, a photograph
          Below, sealing mechanism.

made quickly and simply. The electrodes do, of course, wear and there is a size limit of about five peripheral inches unless quite enormous capacitor banks are used. Pure nickel lids are favoured since they do not require plating and there is therefore less risk of oxidation at the seal interface, but plated NiFe or Steel can also be used. This style of package is shown in figure 7.4.

Seam welding is currently very popular as a sealing method where a thin flat or flanged lid is presented to a 1 mm thick side wall and two rollers pass simultaneously down opposing edges carrying out a number of pulsed welds. The electrode wheels are bevelled so as to give minimum contact. Two designs of machine are in general use, one requiring a flange thickness of .005" and, to give acceptable resistance, cobalt containing NiFe alloy lids. A single capacitor system is used. The other machine is larger and is less restricted in the choice of lid thickness and material, since each roller is supplied by a separate discharge unit. The style of package known as solid side wall is shown in figure 7.5. The Ni or Au plating is applied either to both parts or Au is plated on the body and Ni on lid. These fuse during the process resulting in perhaps more of a braze than a weld. Adequate joints have also been made using a plated lid and an unplated surface on the package.

Other welding techniques can and have been used such as electron beam, cold weld and laser, the latter having great potential due to its ability to accommodate a wide variety of lid materials and variable package geometry. It has the added advantage that the power source is outside the glove-box and the laser beam is projected through a quartz window.

METAL PACKAGES

For over half a century glass-to-metal seals have been employed to provide lead-outs from an evacuated or gas-filled enclosure. This process was pioneered by the manufacturers of thermionic valves and depends on achieving a joint between the glass and an oxide layer on the metal components for success.

Two design options are available.
    (a) Compression Seal

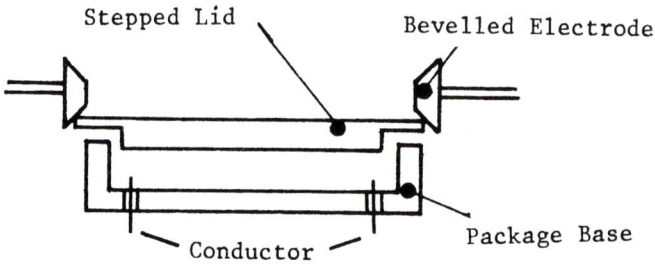

Figure 7.5 A solid sidewall package
Above, a photograph
Below, sealing mechanism.

(b) Matched Seal

The compression seal is made by using metals which do not match the coefficient of expansion of the glass - the outer metal part has a high coefficient, the glass medium coefficient and the centre conductor a low one. During the cooling phase after the firing of the glass, the outer part shrinks onto the glass and the glass, in turn, shrinks onto the conductor.

In the second case all three components have similar expansion coefficients and they are therefore wholly dependent on the molecular joint to the metal oxide for success, since in this case no mechanical assistance is available. For hybrid packages the matched seal is more often used as it withstands temperature cycling better. A number of metals are used each with a different T.C.E. value but in general they are nickel iron alloys or steel, the most popular of the former has a small cobalt content and is widely referred to as Kovar, although there are several direct equivalents of this Westinghouse trade name.

All metal packages have evolved from the small circular TO5, TO3 types of semiconductor enclosures and with time the size has gradually increased. The shape is generally rectangular to conform with the shape of the substrate and two basic constructions are currently in use.

(a) Platform, as shown in figure 7.4
(b) Solid Sidewall, as shown in figure 7.5

The Platform package has pins on either two or four sides, and is enclosed by a raised lid which is soldered or welded into place. Soldering can be achieved using either an overhanging lid, with a preform under the lid flange, or by forming a fillet up the side of a flangeless lid.

The Solid Sidewall package is generally constructed from 1 mm thick material either pressed in one piece or prefabricated from a flat plate brazed to a metal "picture frame". Holes are drilled in either two or four sides of the base to provide dual-in-line or plug-in formats or through the walls to provide flatpack (sometimes called butterfly) or single-in-line constructions. The lid joint is usually made by welding a flanged lid of similar material to the rim of the package or alternatively soldering using transferred heat and a flat lid with a preform (generally referred to as a combo lid).

CERAMIC PACKAGES

Ceramic Packages can be manufactured in three styles.

(a) As a container package
(b) As a multi-layer container package
(c) As an Integral Substrate Package (ISP)

In the first case a ceramic (alumina) base is metallized with lead-out tracks using a refractory metal system, usually MoMn, at a temperature in excess of $1400°C$. A ceramic frame is metallized on its top surface in the same way and is then joined to the base with glass at about $1000°C$ leaving pads exposed on inside and outside of package. Leads are brazed on the outer pads using AgCu at $820°C$ and the assembled substrate is placed inside and is electrically connected to the package by wire bonding. The final operation is to reflow solder a metallized ceramic or metal lid as shown in figure 7.6.

The second approach is a more recent innovation where layers of 'green' unfired ceramic tape are printed with tracks using refractory metal, usually MoW. One of the layers is a ceramic picture frame and the assembly subjected to a single complex sintering operation. Multilayer interconnection can be produced in the base of the package. This could be useful where a circuit can be realized using just chip components that can be bonded directly to the plated, metallized tracks in the base of the package.

The ISP approach uses the printed thick film hybrid as the base of the package and is then treated in one of three possible ways :

(a) A dielectric is printed over the lead-out tracks and a metal layer is printed on the dielectric. A recessed ceramic lid that has been premetallized on its mating surface is soldered to the metal layer.

Packaging of Hybrid Circuits 117

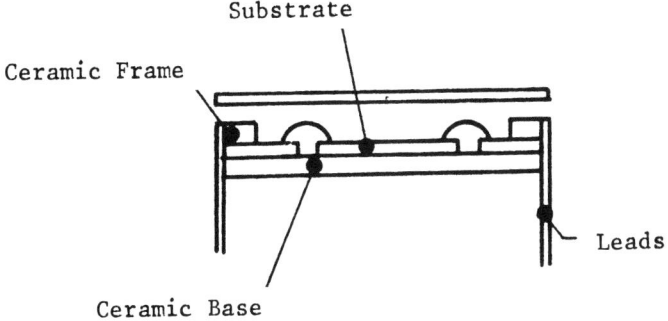

Figure 7.6 Schematic diagram of a ceramic container package.

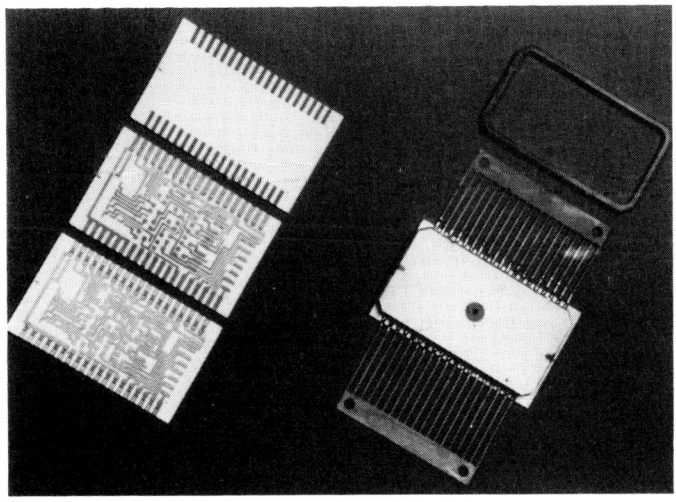

Figure 7.7 Integral substrate package with a flat ceramic lid

Alternatively a metal lid can be used.

(b) A ceramic frame is pre-metallized on one face and pre-glassed on the other and is glass jointed to the substrate in air at about 500°C. After trimming and assembly of components either a metallized ceramic lid or a metal lid is soldered to the top of the frame as shown in figure 7.7.

(c) A metal - usually NiFe - frame is soldered to the metal layer in a similar way to method (a) and a metal lid is welded to this frame.

In all cases the last operation is the soldering of the lead clips. Theoretically, there are a number of advantages in the ISP approach - less weight, less organic content, fewer joints, better thermal transfer to the heatsink, flexibility, and the fact that it is non-magnetic. On the other hand, the entire responsibility for constructing a hermetic package is now in the hands of the hybrid manufacturer which may be stretching his capability. To purchase a tested box that only requires a lid to be added has many attractions.

FINISHING

Metal packages, of whatever construction, require a plating finish for several reasons :

(a) Protection of the body against rust
(b) Assistance with the welding or sealing process
(c) Protection of the pins against rust
(d) Assistance with the bonding or soldering process

Steel and nickel - iron alloys, especially those containing cobalt such as Kovar, are particularly prone to rusting. They are to some extent protected by the oxide present after glass sealing, but this surface is quite unsuitable for any subsequent joining operations. The oxide layer must be removed, usually by dipping in a solution of HCl, and then plated or immersion-tin coated.

A number of alternatives may be used for the plating, such as :

(a) Gold final plating on an electrolytic Nickel undercoat applied to both body and pins.
(b) Electroless Nickel on the body and pins with a gold top-coat on the pins only.
(c) Tin plating on the body and pins.
(d) Tin plating on the body and gold on the pins.
(e) Gold on the body and tin on the pins only.
(f) Nickel on the body and tin on the pins only.

Typical plating thicknesses are :

Gold 1 to 2.5 microns
Electrolytic nickel 0.5 to 0.8 microns
Electroless nickel 5 to 8 microns
Tin 2 to 3 microns

It should be pointed out here that if a package is required to have an earthed pin, then the earth pin must be in the finish selected for the base. This can present problems with soldering, if, for instance, the rest of the pins are gold and the body is electroless nickel which requires an activator for adequate soldering.

QUALITY CONTROL

There are obviously two stages when inspection of the packages takes place.

(a) Before leaving the package manufacturer's plant
(b) After final sealing

The exception to this is the case of non-hermetic or ISP packaging concepts when the responsibility for assembly is in the hands of the hybrid manufacturer.

Package Inspection

The following are the principal inspections carried out :

(a)	Visual	i)	Blemishes in metal or ceramic parts – final plating
		ii)	Quality of glass seal – bubbles, cracks, wetting angles
		iii)	Leads – distortion, twist
(b)	Dimensional	i)	Flatness – base, seal – rim
		ii)	Linear – working height, OD constraint
		iii)	Pin length – internal, external
(c)	Hermeticity	i)	Leak test – failed glass/metal seal
		ii)	Leak test – failed glass/ceramic seal
(d)	Environmental on batch basis or type testing	i)	Temperature cycling
		ii)	Salt mist
		iii)	Lead bending
		iv)	Humidity
		v)	Electrical/Insulation
		vi)	Plating Control

To examine in detail each of these particular controls would be to write a quality manual. These exist for packages in both the suppliers and users factories, each laying stress on those items which have historically given trouble. Leaking seals and plating finishes seem to give the most problems.

Hermeticity is checked at the package manufacturer's plant by placing the package cavity onto a seal ring on the orifice of a mass spectrometer. A pressure differential of nearly 1 atmosphere is created and helium gas is propelled over the package surface. Any gas sucked into the package will be detected by the mass spectrometer. The average leak rate considered unacceptable is approximately 10^{-8} cc per sec per atmosphere (see Mil. St. 202D method 112 condition C). The plating, apart from visual constraints such as pits, burns, blueing, blistering and general discolouration, has to maintain the agreed thickness over the entire surface. Gold thickness is quite easily measured using a betascope. Electroless nickel thicknesses can be

Packaging of Hybrid Circuits

ascertained by weighing or electrical measurement, but the more widely accepted method is the measurement of thickness from a polished section. Electrolytic nickel is yet more difficult to either assess or to control.

FINAL TEST OF PACKAGE

The user will in general have a test programme that ensures that his products meet his customer's requirement but particularly for the package he will carry out the following:

- (a) Visual
  - i) Quality of weld/solder joint
  - ii) Damage to external surfaces
- (b) Hermeticity
  - i) Gross leak - bubble test
  - ii) Fine leak - mass spectrometer

Apart from the need to ensure correct operation of the circuit itself the maintenance of hermeticity and thus the maintenance of the slight internal pressure of the nigrogen filled package is of paramount importance to the hybrid manufacturer.

Gross leaks in a package can be detected by a simple bubble test. To search for fine leaks, the mass spectrometer is again used in this case to detect traces of helium leaking from the package either put inside with the nitrogen or allowed to penetrate in through a faulty seal during an overnight 'bombing' in helium.

PRODUCTION COSTS

In order to assess the relative costs of the package types and the various process variables some guide on estimated production rates could be useful. This of course can only be done in general terms as there are so many variables from one plant to another.

Package Costs

These are placed in order, the most expensive at the head of the list:

- (a) Metal/Glass Flatpack

- (b) Ceramic Flatpack
- (c) Metal/Glass Solid Sidewall DIL
- (d) Multilayer ceramic DIL
- (e) Ceramic I.S.P.
- (f) Metal/Glass Platform DIL
- (g) Ceramic Cover ISP/DIL
- (h) Fibreglass Frame Container Package
- (i) Plastic Cover ISP/DIL
- (j) Conformal Coated Substrate

Sealing Costs

As a guide, the relative speeds of the various lid sealing techniques are given below. However, in the case of welding it is the speed at which the package can be fed to and from the welding source that has more significance. The higher rate operations are given first.

- (a) Laser
- (b) Electron Beam
- (c) Projection Weld
- (d) Dual Discharge Seam Seal
- (e) Single Discharge Seam Seal
- (f) Solder – Preform
- (g) Glassing
- (h) Solder – Hand

Plating

The price of all manufacturers packages using any form of gold will depend on the value of the gold at the time of purchase. The following comparisons are made but should be viewed as a guide.

A 20-pin platform package base of approximately 1 square inch area is used as the basis for calculation. The prices are relative to the cost of the package with gold at $450/oz.

Cost of unplated base	100
Nickel .5 –.8 microns thick plus gold overall 2 microns thick	88

Nickel .5 - .8 microns thick plus gold overall 1.3 microns thick	67
Nickel 3 - 5 microns thick plus gold on pins only 2 microns thick	50
Nickel 3 - 5 microns thick plus gold on pins only 1.3 microns thick	42
Tin 3 - 5 microns thickness overall	17

THE FUTURE

More and more devices will be packaged in chip carriers of one sort or another and thus an increasing number of hybrids will require some form of less costly overall protection. Activity is however present at either end of the size scale. A number of closely packed chips can be put onto a substrate with interconnection and then enclosed in a conventional small DIL or flat package. There is also an attraction in making very large hybrids and putting these and other components into one module package since this offers easy field replacement and better security. In microwave circuitry more chip components are being used and hence there is a need to use hermetic packaging. Fibre optic transmitters and receivers are tending to have the laser diode and control hybrid in the one sealed package and integrated SAW hybrids require hermeticity. Finally there will be a continuing requirement for packages with irregular geometry and in unusual materials.

CONCLUSION

The packaging of Hybrid Circuits is not easy, can be expensive and can add much time to the eventual delivery of the circuit. The best advice is to plan early, persuade the customer to work within a narrow range of sizes, to be quite certain of the customer's environmental and mechanical requirements and work closely with a package manufacturer.

8
Automating the Production of Hybrid Circuits

G. W. GRIFFITHS

INTRODUCTION

The manufacturing methods used for the production of hybrid circuits will depend to a large extent on the end use of the product. For thick film, as an example, we can find diverse applications; from mass production car electronics, to very small series for space equipment with a nearly continuous spectrum between these extremes. Although the principles of printing, firing and trimming may be common, the production methods, quality checks, batch control and other factors will be very different. The important matter of attaching active and passive devices, together with the final packaging will have an even wider divergence of methods, to a point where there is little in common between the products. The production will vary from fully automated re-flow solder techniques for large series consumer and subscriber type telecommunications applications to clean room assembly methods using semiconductor chips. Hand assembly and automated bonding, controlled by micro-computer, are used for small and medium quantity production of high reliability circuits for trunk system telecommunication, military, aerospace and similar applications.

The majority of purchases of circuits from a specialist hybrid house will be made by the users of medium quantities of circuits for industrial and high reliability applications. The term industrial is used to mean a rugged, reasonably priced device with a small probability of failure under not too severe conditions for 5 years or more. "High reliability" is less clearly defined but means that all known failure mechanisms must be either avoided or accepted only if their effect can be seen in the very long term.

The remainder of this chapter will be devoted to a description of production methods, automation in practice and the future possibilities for further automation in the area of industrial and high reliability hybrids in the medium quantity range. One of the advantages of the thick film method is that it can be automated, and a number of high and very high volume applications exist. However the techniques used in these applications are of a proprietary nature and therefore beyond the scope of this chapter. It is assumed that the basic technology is understood, but where a specific relationship between the technology and the automation is relevant it will be emphasised.

PRINTING, DRYING AND FIRING

The first processes are those forming a sequence of printing, drying and firing, there being up to 10 printing and drying operations and rather less firing operations. A first stage of automation is by using multiple circuits on one standard sized substrate, and leaving them as such until as late in the processing as possible, at which point they are laser scribed and separated. Prescribed substrates can be used, but require a prior knowledge of the disposition of the circuit on the substrate giving rise to stock holding and control problems and reduced flexibility. While it is clear that wet substrates cannot be stacked, it may also be undesirable to stack dried unfired substrates (it is not so often appreciated that small scratches on fired resistors can have an adverse effect on long term stability). Consequently it is normal to use a cassette system whereby each substrate is separate and held only at the edges remote from critical circuit depositions or components. Once the concept of multiple circuits on standard substrates is accepted, then the need for flexibility in the substrate processing area is largely avoided, as all substrates will have the same dimensions. Removal by indexing the cassette vertically and pushing the substrates horizontally on to the moving belts of the dryers and furnaces is possible. A similar method may be used for loading the shuttle or rotary table feeding the printers. As there is a need for altering belt speeds, removal from the belt and back into the cassette is by sensing the presence of a substrate. A variety of means are possible but simple optical methods are adequate and reliable.

TRIMMING

The resistor trimming stage has been automated for a long time and all the resistors on a substrate may be trimmed at a single loading. The use of computer control has been common practice since laser trimming was introduced. Probe connections, steering of the beam, type of trim etc., are all achieved by programming a mini or micro computer. Loading the substrates by hand can take more time than the actual operation of passive trimming which at the rate of 15,000 resistors/hour and 15 resistors per circuit means a capability of 1,000 circuits/hour or about 1 every 3.5 seconds. Substrates with multiple circuits are trimmed using a step and repeat table on the substrate holder and most laser trimmers offer such facilities. Entry into the trimming position with either rotary or slide mechanisms can be the same as is used for printing. It should be mentioned however that active or functional trimming usually takes considerably longer - in some cases minutes per circuit and is usually carried out just prior to the final packaging operation. It is not practical to automate the handling unless the production volume of any one part is sufficiently large.

ACCURACY OF EQUIPMENT

For the above, and to a large extent for the assembly operations (with the exception of soldering) the whole system needs to work not to a relative accuracy, but to an absolute accuracy. Each pattern, from the artwork stage on, must be kept within limits that will ultimately relate to two edges of the substrate. All machines must relate to each other such that when a new screen is fitted to a machine no adjustment of position should be needed. This necessitates a system of preparation of artwork, photography and screen-making that involves jigging to accuracies that are probably not appreciated from a casual study of the technology, but are vital if any kind of mechanisation is to be successful.

ATTACHMENT

The substrates now have the components attached. Although

some mixed technologies are used it can be assumed that these will
incorporate methods drawn from the two main ones which are (a) chip
and wire plus chip capacitors, using epoxy attachment methods,
and (b) reflow solder methods using pre-encapsulated semiconductors
and chip capacitors, but this time soldered.

The chip and wire methods can have two subdivisions, one
where the hybrid will be hermetically sealed, and the other where
external connections will be by means of soldered lead frames
attached after the added components have been protected.

Die Attach

The epoxy used to attach devices can be printed using the
standard methods for thick film. The semi-conductors are obtained
as slices and after sawing are presented to the chip attach machine
as a separated but ordered array of dice on a tacky foil. Removal
from this foil is by manual alignment over a needle, using closed
circuit television (CCTV). The needle rises, and from the underside pushes the dice away from the foil and into contact with a
capillary tube connected to a vacuum system. The tube then conveys
the dice to the correct place on the substrate. With the use of
CCTV, pattern recognition is possible and has lead to the use of
some self aligning machines on dice pick-up for semiconductor
manufacture. The possibility of using this for hybrids is obvious and
is receiving attention and hence it can be expected that at some
time, dice placement could be fully automated. A system of X-Y
controls pre-set to the needs of specific circuits is used. These pre-
set positions are now determined by micro-processor controlled
positioning since this is the cheapest and most flexible method as
well as being highly adaptable, with easy software storage. Again
the loading of substrates can be by means of cassettes. Several
passes may be needed to accommodate a range of semiconductor
devices. For some rather special IC types, the so called waffle
trays are used, in which the IC's are supposedly pre-tested and only
good ones provided by the IC manufacturers. This method is used
where quantities are small or slices unobtainable. It is not
considered to be a good candidate for mechanisation as there is
only a 1 in 4 chance that the orientation is correct for any one
dice.

Wire Bonding

For wire bonding much has been done in the semiconductor industry to minimise costs, and the hybrid manufacturer naturally looks to see what is relevant here for multiple chip bonding (apart from sending circuits out to the Far East). The precision of placement needed for totally automatic bonding cannot be met with multiple chips on hybrids, so any method must allow for some degree of variation in die position. The type of automation that has been applied to multiple transistor circuits has so far been rather primitive, automatic movement from one chip to another is possible but the actual bonding is by hand control. Automation, or at least machine control, is essential for multiple IC circuits of the complexity now being seen. Where a circuit has more than 300 wires, then the chances of mistakes or omissions are high if the bonding is done by hand while looking at a diagram. The checking will take nearly as long as the bonding and is itself subject to error.

Machines exist which can be programmed to do this bonding, with the final positional accuracy done by hand. A microprocessor controlled system deals with the sequencing and memory of the bonding pattern.

For programming, an example of the circuit is placed on the machine which has a spot light corresponding to the bonding point. Using manual control of the substrate table, two clearly identifiable points, as near to diagonally placed as possible, are found on the substrate metallisation pattern, and the positions are stored. The first IC is located similarly by two reference points. The wire bonding routine for this IC is now done with each bond being referenced to the datum points. This process is repeated for all other IC's. In the case of transistors an absolute position may be defined. For the purposes of machine setting the header is treated as an IC. When the pattern is checked and found satisfactory, a copy of the wiring schedule and positions is taken on disc. It could be argued that all this might be done off the machine, and in theory this is possible, but in practice there are so many little things such as loop control of small bond lengths etc., that this "walk through" method is much preferred.

When the bonding of circuits starts, the machine goes in sequence to all those points defined as references and waits for a correction to be made. When all points are corrected the "go" button is pressed and all the bonds are made.

It can be seen that such a machine, whilst giving a saving of up to two-thirds of the time over manual bonding, does much more. Since installation yields have increased due to avoidance of errors, operator training is reduced and consistency is guaranteed. More recent machines offer pattern recognition to perform the alignment corrections. Figures 8.1, 8.2 and 8.3 show three different machines. These machines are progressively more sophisticated and demonstrate the progression of facilities available. A typical circuit might take 12 minutes on the manual machine, 3-4 minutes on the manually aligned automatic bonding machine and $1\frac{1}{2}$-2 minutes on the fully automatic machine.

Reflow Soldering and Surface Mounting

Where size is not critical and components can be soldered, then a quite different method can be used - that of reflow soldering.

Reflow soldering for hybrids is a method whereby solder paste is applied by screen printing and the areas to be soldered are defined by solder dams. These dams can be either dielectric or resistor material over-printed across conductors to prevent flow of solder beyond them. The components are designed for surface mounting and are placed, with no need for great accuracy, on to the solder printed areas. The whole substrate is then heated under the control of a "temperature/time" sequence on a belt to melt the solder and thus connect the parts. Surface tension will pull the parts into the designed positions, hence the lack of need for precision. Parts may be bowl fed, rack fed or simply put on by hand. One of the advantages of surface mounting is that it is an extremely fast operation and with a reasonable mix of components, up to 4,000 per hour may be placed. Ideas on component assembly methods deriving from the more traditional origins can be abandoned in many cases. This method has now attracted the attention of both machine builders and component suppliers, and much activity is to be seen in position-controlled placement machines where the components are presented in an orderly manner, using tapes, vibratory bowl feeders etc..

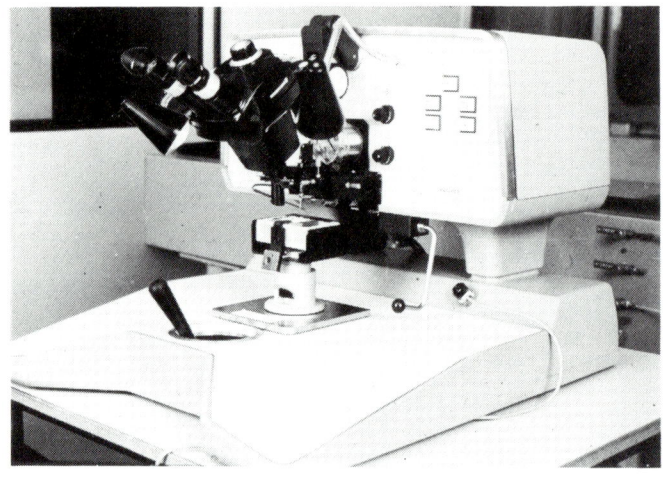

Figure 8.1　A manually aligned manually operated wire bonder.

Figure 8.2　A manually aligned automatic wire bonder.

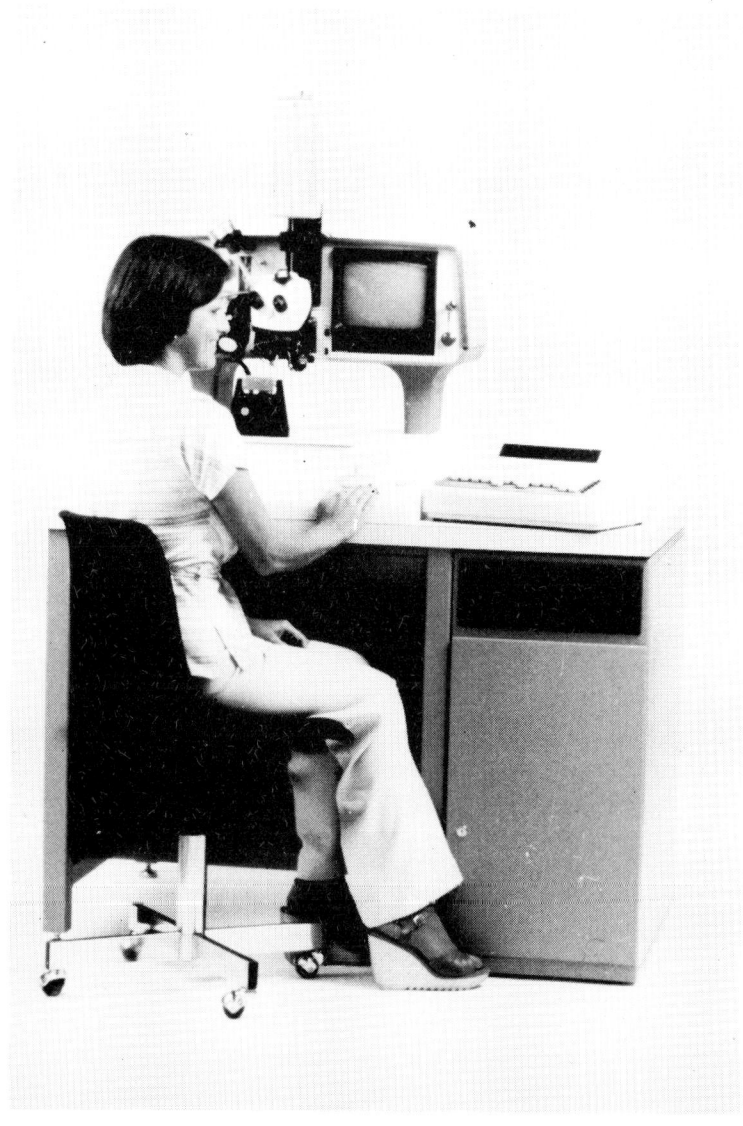

Figure 8.3   A fully automatic wire bonder employing pattern recognition to determine the die positions.

The realisation by the component manufacturer that some better degree of automation is needed within the whole electronic assembly industry has lead to the introduction of components specifically suited to surface mounting. The use of these is possible on many types of "substrate" and as a consequence looks like supplanting the wire ended, inserted component. The types of X-Y placement machines developed for P. C. assembly of surface mounted components now have a programming and set-up flexibility well exceeding that using drilled boards and bandoleered wire ended components. The use of such machines, adapted maybe to the generally smaller hybrids, is obvious. In this area it is becoming irrelevant to distinguish between hybrid and P. C. B.. Various considerations may lead to the choice of one substrate material or another, although the merits of the use of these materials is another matter. What is relevant is that common methods of assembly can be adopted. A prime consideration for mechanisation for surface mounting is that most surface mounting components are not identifiable for value, due to their small size and construction. Any mistaken placements are not correctable by visual inspection and could have a disastrous effect on the product. Emphasis must be placed on meticulous material control - preferably by suppliers providing taped components (or some other orderly method not requiring user re-sorting) and totally mechanised pre-determination of the positional placement. So far, the much discussed chip carrier has eluded such treatment, but must come in if it is to form any significant impact on the use of higher levels of silicon integration in medium or large scale electronic assembly. The acknowledgement of the reliability of modern silicon devices, and a realistic life of use for equipment, has lead to an acceptance of the methods described above for assembly of many telecommunication products, which must considerably enhance the role of surface mounting in the electronic scene in general.

The summary of this situation is that for the manufacturer in the medium series quantity range, re-flow soldering offers a potentially good solution for economically priced industrial hybrids.

SUBSTRATE SCRIBING

For the hermetic hybrids the substrates are "scribed" using a CO_2 laser under the control of a micro computer just prior to insertion into the package. This is a simple device with thumb wheel switch controls and can be reset from one cutting array to another in 2 to 3 minutes. Scribing takes only a few seconds per substrate. In the case of substrates with clipped on lead frames, the scribing is carried out just prior to the lead frame attachment.

PACKAGING AND TERMINATION

To make the hybrid accessible to the user, some form of a termination must be provided. For hermetic types this is part of the wire bonding. For others it is usual to apply clip-on leads (originating as reels) with the single sided or double sided connections being made by pushing on the leads and soldering. Prior to this, any connection that will be omitted from the final circuit is cropped off. The lead frame is also cropped after insertion into manageable lengths for solder-dip and defluxing as multiple units. Figure 8.4 shows a machine for the automatic attachment of the lead frames, and figure 8.5 shows the solder dipping of multiple substrates to attach the lead frames to the substrates. The plastic encapsulation by liquid or fluid bed will also use these multiple units.

For the psuedo-hermetic circuit with glued-on ceramic lids, these lids will be attached before the terminations to protect the naked semiconductors from solder and flux. Hermetic circuits are sealed in a closely controlled dry nitrogen atmosphere by a method of high current projection welding. The speed of this operation is so much faster than all preceding ones that the main consideration is prevention of mechanical damage to the contents rather than the ultimate in speed. Seam welding may also be used, offering a higher degree of tooling flexibility at the cost of process time.

TESTING

100% testing is essential and must be considered part of the process. With such a variety of end uses the parameters of testing seem limitless. For hybrid circuits only the outside terminals are available, although to a limited extent in-line tests can, and in some cases must, be made. Testing is carried out using a mini-computer controlled tester with full d.c. test and a.c. up to 100kHz. High frequency a.c. tests may be made using interface

Figure 8.4   Clipping the lead frames on to hybrid substrates.

Figure 8.5   Solder dipping multiple units of hybrid circuits to attach the lead frames.

boards and IEEE 488 bus-addressable test equipment.

CONCLUSION

The preceding shows some of the existing ideas on automation. It is also hoped that the need for the maximum flexibility in both thought and practice has been demonstrated. What a few years ago was believed to be the ultimate in complexity is now the norm. The future will be in several directions, one at least will be that of ever increasing complexity, and the adoption of self learning machines with larger memories etc. must be expected. On the other hand many hybrids will remain simple, the future for these lies with good process control and effective but flexible automation.

Over the whole range of hybrid manufacture the use of computer aided operations is seen from CAD in design to computer controlled test. Many of the systems, CAD, trimmers, wire bonders etc., are of the "Turnkey" nature whose operating system are virtually inaccessible to the user, making interactive programming impossible. Some trends to correct this can be seen to be taking place and it is the hope, and expectation, that the CAD design work will provide much more than just the basic photoplotting. Trimming programmes, wire bonding etc. and, not least, a word processing facility to provide the control documents to see the product safely and correctly through the factory can in theory be derived from the initial design. Only in one case has the full power of the computer been seen to complete the job. For surface mounting at the moment one machine will accept programming in any order and execute a least path sorting routine to speed up the operation. There is a long way to go but a more definite route to take can now be seen.

9
The Application of Hybrid Techniques

N. G. BURROW

INTRODUCTION

The advantages and disadvantages of the various films, substrates, add-on components and packages have been discussed elsewhere and in this chapter it is intended to bring these together to show how they impinge on actual hybrid applications without unnecessary repetition of earlier discussions. In the author's view the feature of hybrid technology which makes it so attractive to such a multitude of users and in a similar multitude of environments is its flexibility. This is demonstrated in its ability to bring together many technologies on a single substrate; the capability of integrating both the control and the power functions on a single substrate; the range of quality specifications (and prices) which is possible; the variety of shapes, sizes and packages that can be used and not least of all relative ease of automation.

In spite of the overwhelming advantages to be gained from the use of hybrids in so many applications, the technology must not be considered as a panacea for all the system designer's problems. For example, if the only obvious advantage of a hybrid for a particular application is its small size, a lot of thought should be given to the alternatives such as squeezing up the PCB or reducing the size of neighbouring circuits before any commitment is made. Similarly, if many large add-on components such as electrolytic capacitors are essential, then hybrids are unlikely to offer advantages over PCB's. This means that continuous liaison between the manufacturer and the user from the moment that the thought of using a hybrid comes to mind is vital if an application is to be successful. The worst possible situation occurs when the user hands over a complete discrete circuit to the hybrid manufacturer with the instruction to "make that" and then leaves him to it. The designer should be willing to adapt his design to the advantageous features of hybrids, where applicable should take into account the current availability of surface mounting devices and should tolerance his circuit completely. For his part the manufacturer should make sure that the user is aware of what is possible and what is difficult, what is preferred and what is to be avoided, what is cheap to do and what is expensive. The willingness to share confidences between the user and the manufacturer is the best possible omen for a successful hybrid.

In the discussion that follows, some of the more sophisticated application areas of hybrids will be described, and in each area the noteworthy features hybrids will be highlighted. Most of the applications exploit more than one advantage of the technology and, in some, a single alternative technology is unthinkable. It must be appreciated that in addition to these specialised applications there exists an enormous number of "bread and butter" industrial, military and consumer-type applications for which, in many cases, the only reasons for using a hybrid are the convenience and cost advantages of a modular system.

Figure 9.1 shows a DIL resistor array, one of the simplest possible film circuits of all and yet the ability to produce this in a form compatible with the DIL outline of silicon integrated circuits is an attraction to equipment designers that must be appreciated. The ruggedness, low cost high performance modular component saves on interconnect and space compared to its discrete counterpart and, within reason, a variety of accurately adjusted resistor values are possible.

At the other end of the scale, figures 9.2 and 9.3 show two examples of hybrids designed to achieve a very high packing density indeed, without resorting to the complexity of chip and wire hybrids. The circuits combine semicustom integrated circuits packaged in single layer chip carriers, SO packaged standard devices, chip capacitors, multilayer interconnect, laser trimmed resistors and the use of several surfaces. In figure 9.2 the hybrid

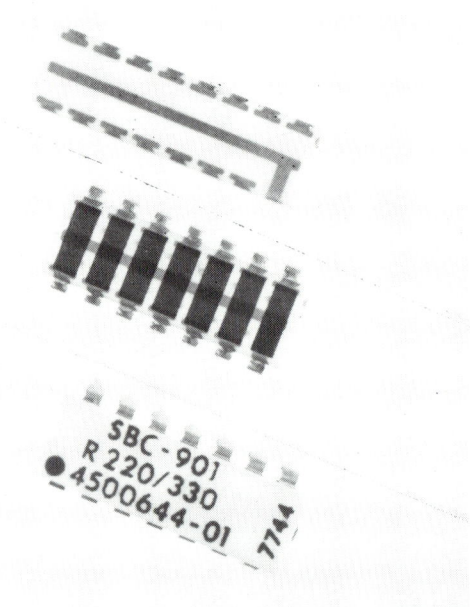

Figure 9.1   A 14 pin DIP resistor array

Figure 9.2   An industrial grade hybrid combining a variety of
package styles and interconnection methods.
(courtesy CorinTech Ltd.)

Application of Hybrid Technology 139

Figure 9.3   A hybrid assembly using stacked substrates, a semi-custom integrated circuit and a custom designed lead frame to facilitate easy interconnection between the substrates
(courtesy CorinTech Ltd.)

Figure 9.4   Voltage regulator hybrids for automobiles
(courtesy Philcom Electronics Ltd.)

is two substrates "back to back" held together by a wide throated lead frame and in figure 9.3 the hybrid uses both surfaces of two substrates held by a custom designed lead frame. The technical challenge of the assembly shown in figure 9.3 is discussed in more detail in reference 13.

AUTOMOTIVE

Hybrid application areas in motor vehicles fall into three principal categories: consumer-type products, automotive functions and transducers. The criteria for the choice of a hybrid technology for consumer-type applications such as radios and burglar alarms are similar to those for other mass-produced consumer goods, the most significant being cost, size and reliability. Modern vehicles require increasing amounts of electronic circuitry to be housed in the ever diminishing spaces remaining in ergonomically designed interiors; hybrids often fit the space and the bill. Reliability is rather more critical than in, say, domestic appliances because of temperature extremes, vibration and shock. Thick-film hybrids are usually twice as reliable as PCB's in such environments due to the smaller number of soldered connections (1). Applications that have already been hybridized include the tuner in AM radios, the IF amplifiers in AM and FM radios, cassette player electronics, stereo decoders and traffic information systems.

In the area of exclusively automotive functions the specifications are quite different from normal consumer applications because of the extremely hostile environment for electronics. Traditionally most automotive control functions have been executed using mechanical or electromechanical systems, and electronic replacements have been resisted largely on grounds of cost and reliability. In spite of the forbidding environment and the natural conservatism of motor manufacturers the situation is now changing very rapidly because it is accepted that thick-film hybrids are supremely reliable in such environments as well as being relatively cheap. This change is being accelerated by safety and pollution legislation as well as requirements for improved performance. Already hybrids have been successfully used for anti-lock braking systems, voltage regulators, contactless ignition systems and fuel injection controllers. There is no doubt that there will be an expansion in the use of hybrids in automotive applications partly due to their reliability and partly due to the particular virtue of hybrids of combining the control and the power functions on a single substrate. An indication of the hostility of the environment is given in table 9.1.

In anti-lock braking systems an electronic approach is necessary because the system must adapt to varying vehicle and road parameters. The weight of the vehicle, for example, may vary from 5 to 40 tons and it must be stopped on dry roads, water-logged roads and ice covered roads. Environmental and performance considerations lead to the choice of thick-film hybrids for this application rather than PCB's. One manufacturer states that his eventual hybrid was only a quarter of the size of an equivalent PCB (2), and consisted of a 3 3/8" x 3 3/8" substrate having 7 films (2 conductor, 2 insulator, 3 resistor) and carrying 100 resistors plus 60 discrete devices including semi-custom IC's. The resistors were initially deterministically trimmed by a laser using the component pads for the measurement contacts. In order to avoid the use of costly and less reliable potentiometers, some resistors were then functionally trimmed during the testing of the complete hybrid. Hybrid voltage regulators are used in what is probably the worst environment in which mass produced hybrids are ever found. Because of their small size they are often mounted inside the alternator casing where thermal cycling and vibration are severe. Nevertheless, one manufacturer reports from experience of two million such regulators fitted in cars that there was an order of magnitude improvement in reliability over the conventional mechanical regulator (3). Figure 9.4 shows three implementations of the same regulator circuit made by Philcom Electronics. The left hand version is a conventional PCB assembly. The middle one is a thick-film hybrid using conventional discrete components. The film resistors are deterministically trimmed before the discrete components are mounted and this is followed by functional trimming to adjust the cut-in voltage of the regulator. The right hand circuit uses a PCB to mount the discrete components and a thick-film resistor network. This is probably easier to assemble where conventional discrete components are used, but most of the components could now be replaced with surface mounting versions making the thick-film hybrid more attractive.

TABLE 9.1

Environmental Factors

Temperature extremes	-50°C to 130°C
Temperature gradient	up to 100°C/minute
Humidity	up to 100%
Contamination	oil, water, acid, solvents, etc
Mechanical	inexperienced manhandling, accumulated grime, road gravel bombardment, etc, severe and varied vibration

Electrical Environment

Complex electromagnetic fields
Interaction between electrical functions
Inductive switching transients on the power line
Variation in supply voltage from 4.5 V to 24 V (for emergency starting)

Application of Hybrid Technology   143

Figure 9.5 Switched resistor network for motor speed control
(courtesy Citec Ltd.)

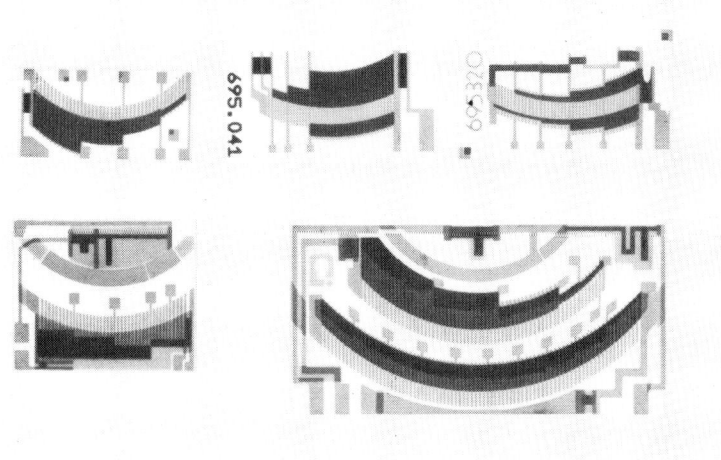

Figure 9.6 Level sensors for automotive fuel tanks
(courtesy Citec Ltd.)

TRANSDUCERS

This is an application area for hybrids which has a great potential for exploitation in the future. Whenever possible in microprocessor control systems the transducers (or sensors) should be electronic in concept, small, of high sensitivity, cheap and reliable. Hybrids offer the particular advantage that in addition to the integrated sensing devices, some of the processing and interface circuits can be included on the substrate.

Many transducers are based on temperature sensing, and for this purpose a thermistor thick-film paste containing spinel-type semiconductor oxides, resistive oxides (RuO_2) and a glass binder is often used. The thermistors are used either for temperature measurement directly or for temperature compensation of other elements. Their reliability is as high as that of bead-type thermistors (4). Radiant heat sensors have been reported which consist of two thick-film thermistors, one covered with aluminium foil (low emissivity) and the other with carbon (high emissivity). An air velocity sensor consists of a thick-film thermistor printed on a ceramic cylinder which is placed in the air flow. A similar film composition is used in humidity sensors, though a new film material containing $MnWO_4$ is claimed to have exceptional properties including a very high sensitivity (7.8%/%rh) (5). A film material containing $LaNiO_3$ is sensitive to alcohol gas and is used in commercial gas detectors.

The well known piezoresistive property of thick-film resistors can be exploited in many transducers as well as in simple strain sensors. The best resistive component is $Bi_2Ru_2O_7$ which produces thick films with gauge factors five times greater than those of conventional metal films, but for which the temperature compensation problems are more severe. Standard alumina substrates are good both because of their high thermal conductivity and because the deformation is elastic almost up to the fracture limit. Piezoresistive transducers are used for the measurement of oil pressure in car and truck engines where they have to respond to pressures up to 10 bars with overpressures of 30 bars and explosion protection up to 100 bars (6). Because they are mounted on the engine they have to withstand a severe environment and normally operate at an oil temperature of about 130°C. Nevertheless, they are reported to be more reliable than conventional transducers in which a metal diaphragm moves a wiper along a potentiometer track.

Both thin and thick films are used in the thermal printheads of printers that have become popular for use with calculators and home computers in recent years. These normally consist of thick-film resistor heater elements which are addressed by orthogonal arrays of conductors. Because of the fine dimensions required for very high resolution printheads (3 mil lines on 5 mil centres for a 200 lines-per-inch printer) etched thin films of gold and nichrome are often used for the bottom conductor layer (7). The dielectric layer and the top layer are usually thick film. The resistor layer has to have rather special properties including a pulse power handling capability of 20 mW/mil^2, a smooth surface and good abrasive wear resistance (because it is always in moving contact with the paper). The heater elements cause a chemical reaction in the paper at about 400°C, the temperature excursion normally being about 200°C and occuring at a frequency as high as 1000 cycles per second. Research is in progress on printheads which use thick films only, though there are problems with uniformity over a 9" x 3" substrate and, of course, the problem of line definition is crucial. In the future it is hoped to use glass or porcelained steel substrates which are potentially cheaper and flatter than alumina.

Thick-film resistor networks are often used in conjunction with switched or wiping contacts to make variable resistors. The network shown in figure 9.5 is manufactured by Citec for motor speed control applications where the power loading is up to 20 W. This particular network is, for example, used to control the fan speed in the heater of the Ford Cortina. The planar type construction with the thick alumina substrate doubling as a resistor base and a switch plate is cheaper, more compact and more reliable than the alternative of several wire-wound resistors and a separate switch. Because of the high currents involved, rivet terminations are used which are wiped on the reverse side to select the required resistance. Custom designed networks are offered with power dissipations up to 50 W, resistances in the range 0.47Ω to 1000 MΩ and voltage ratings up to 25 kV. For lower current applications the conductor films are wiped directly. The thick film sensors shown in figure 9.6 are used as the level sensors in automotive fuel tank senders. A metal wiper is moved by a float in the petrol tank to set up a resistance corresponding to the fuel level. The varying resistance is used to control the visual display of a simple current meter. A considerable advantage in using thick-films for this application is that the resistor law of the track can be tailored to the dimensions of the tank to give a linear reading of fuel level.

MEDICAL

Thick-film hybrid circuits are used in hearing aids, vision aids for blind people, cardiac pacemakers, through-the-body FM telemetry systems and external capacitive biomedical probes for ECG and EMG measurements. Since some of these applications are, in fact, life support systems reliability is the most important single feature of hybrids which justifies their use. Apart from the obvious considerations of size and weight the following advantages of hybrids are also of interest :

> Wide frequency range - hybrids allow low frequency pulse circuits and RF telemetry circuits to be integrated on the same substrate.
> Easy combination of analogue and digital circuits.
> The capability to use state-of-the-art devices of any technology for the utmost in performance, size and weight.
> Layout flexibility - the hybrid can be shaped to a physiologically acceptable configuration.
> Hermetic packaging - necessary to avoid any possibility of body fluid contamination.

The most important implanted application is in cardiac pacemakers which are used to correct rhythm disturbances following open-heart surgery and congenital disorders affecting the rhythm of the heart (8). Early pacemakers consisted of simple blocking oscillator circuits containing only 10-15 discrete components. They could not respond to the actual activity of the heart so they could potentially be in competition with it. Ideally, of course, the pacemaker should stimulate the heart only when such stimulation is required and this demands rather more complex circuits. State-of-the-art pacemakers incorporate both double heartchamber sensing and double heart-pacing. In the future they will be even more sophisticated and will be required to monitor pH levels, respiration rate versus heart rate, arterial pressure and various cardiac electrical activities. In spite of this increase in complexity the size and weight of implanted pacemakers continues to fall. The projected changes in weight and diameter from 1970 to 1990 are : weight, 200g to 40g; diameter, 7 cm to 3 cm.

There is a growing requirement in the medical profession for miniature FM telemetry systems to transmit signals either to or from the body. Some implanted systems are needed only intermittently so it is desirable to maximize the life of the batteries by switching them off when not in use. This is done by coupling an RF signal from an external transmitter to an aerial on the system. An amplifier and a demodulator then provide the switching signal. Obviously the package, although hermetic, must be non-magnetic so a welded stainless-steel package is normally used. Telemetry signals in the opposite direction are used to monitor ECG, EMG and other physiological data. Similar systems are used for the biotelemetry and radiotracking of wild animals where biologists try to limit the weight of the devices to 2% of the animal's weight (9). In these applications unencapsulated chips are sometimes used because of the size and weight benefit. The hybrids are usually coated with silicone.

Hybrids are used in hearing aids of both the "body worn" type and the "behind the ear" type. Obviously the main criterion is miniaturization, but low noise considerations are also relevant. In some designs of "behind the ear" aids the circuits and the batteries are contained in a sealed case which is regarded as a throw away unit. The "Sonicguide" blind aid uses ultrasound to identify objects and their distances in a similar manner to radar. A transmitter, two transducers, two receivers and two earphones are all mounted in a special spectacle frame which is worn by the blind person. Developments in artificial vision, on the other hand, use implanted hybrid circuits for stimulating the brain of the blind person. In one development, for example, sixteen thick-film hybrid circuits were fitted to the cranial implant.

ACTIVE FILTERS

The applications of hybrid active filters are as wide as the spectrum of uses of filters and include industrial, communication and military. Early hybrids used thin-film technology, but nowadays thick films are used almost exclusively as a result of improvements in materials. The advantageous features of hybrids for active filters are :

Good reliability
Good stability
Good repeatability
Low circuit strays
Easy custom design
Pretunable (before packaging)
Low temperature dependence ($\pm$ 30 ppm drift in f_o)
Variety of packages
Low price/performance ratio.

Many hybrid houses offer quick custom and semi-custom design services. The basic circuit configurations are well tried and design rules are simple and reliable. Non-critical chip capacitors are selected which allow all resistors to be printed with a single paste so that the TCR is consistent over the whole temperature range. An example of a fifth order hybrid filter made by Newmarket Microsystems is shown in figure 9.7. This is an elliptic bandpass filter having a passband attenutation of 0dB (with < 1 dB ripple) extending from 50 kHz to 100 kHz. The stopband has better than 55 dB attenutation for frequencies less than 35 kHz and greater than 125 kHz. It is mounted on a 1" x $1\frac{3}{4}$" substrate and hermetically sealed in a 34-pin DIL metal encapsulation. Other encapsulations which are available are : phenolic resin, epoxy, ceramic case and metal flat pack.

"Universal" hybrid active filters are extremely useful building blocks for an enormous number of applications. They are usually based on 3-amplifier bi-quad designs which can implement low-pass, high-pass, band-pass, notch and all-pass functions. Usually the LP, HP and BP functions are available simultaneously for use, for example, in loudspeaker crossover units, and sometimes an additional amplifier is included which can be used to generate complex poles or zeros and to set the overall gain level. Typically the hybrids contain an unencapsulated quad op-amp chip (to reduce strays), two capacitors and up to ten thick-film resistors. The

Application of Hybrid Technology   149

Figure 9.7   Fifth-order active filter
                     (courtesy Newmarket Microsystems Ltd.)

Figure 9.8   A 10 megabyte per second fibre optic receiver
             unit   (courtesy Belling Lee Ltd.)

centre frequency, the Q and the gain (if necessary) are defined by external resistors and the filters may be cascaded to produce higher-order responses. The following specification is typical of a state-of-the-art universal second order hybrid filter by Newmarket Microsystems.

Functions:	HP, LP, BP and notch (simultaneously)
Types:	Chebyshev (with two external resistors)
	Butterworth (with three external resistors)
Frequency range:	dc to 150 kHz
Max Q:	500 dc to 1 kHz, down to 1 at 150 kHz
Max Power:	200 mW
f_o stability:	± 100 ppm
f_o repeatability:	± 1%

TELECOMMUNICATION AND HIGH FREQUENCY

The telecommunications industry is probably the biggest single applications area for hybrids and is likely to remain so for the foreseeable future. Thick-film hybrids are used in electronic telephone exchanges in several countries including Sweden, Germany, France and Great Britain. British Telecom uses numerous hybrid circuits per customer line (10). Ericsson of Sweden has developed a one-piece electronic telephone consisting of a hybrid amplifier with an electret microphone attached, an acoustic filter and a case. The speech transmission circuit uses a double sided hybrid consisting of two ceramic substrates joined back to back and interconnected around the edges.

Modular hybrid transmitters and receivers for 10 MHz fibre optic systems are already available. One of the benefits of the hybrid technology is that the two critical front-end devices in the receiver, the PIN photodiode and the FET preamplifier, can be mounted very close to one another on a high quality substrate material. This has the advantage of enhancing the receiver bandwidth as a result of the reduced capacitive loading on the PIN photodiode, as well as reducing the receiver noise. The small size, reliability and functional trimming capability of the technology are also exploited. Figure 9.8 shows such a module by Belling-Lee.

Inductors present a problem in all integrated technologies. However, planar inductors can be fabricated in both thin-film and thick-film technologies and are suitable for applications in the 50-200 MHz range. The maximum practical inductance is about 1 μH and the maximum Q is about 30. At low frequencies the Q is too small to be useful and at higher frequencies into the microwave region the physical size becomes too small to fabricate. Thick-film transformers using planar coils have found application in hybrid flyback isolation amplifiers where they are considerably less bulky than wirewound transformers and have excellent isolation characteristics; the common-mode voltage rating of the isolators is 8 kV and the common-mode-rejection-ratio is 120 dB (11). The input and output windings each consist of three coil layers of eight turns each on separate back-to-back substrates. The substrates have a hole on the middle of the coils for a ferrite core. Although the fundamental frequency of the carrier is only 200 kHz much of the energy of the power pulses extends to several MHz so a wide bandwidth is necessary. The isolator has a flat response from about 10 kHz to 10 MHz which is achieved by using very closely spaced coils of only 5 mils line width and 7 mils spacing.

Hybrid microstrip circuits have become popular for low frequency (1 - 40 GHz) microwave applications because of their cost advantage compared to waveguides, their small size and their compatibility with modern semiconductor devices which can operate at these frequencies. The microstrip circuit consists of a substrate with a gold plated ground plane on the bottom and a microstrip line on the top. The impedance and the wave velocity depend upon the thickness of the substrate, its permittivity and the strip width. Nearly all passive microwave functions can be realised and active circuits can be made with such add-on devices as varactors, switching diodes, impatt and trapatt diodes and bipolar transistors. These are usually in chip form and are wire bonded. The critical component is the substrate which must have a fine surface finish, low loss, high permittivity and low temperature dependance of permittivity. Alumina is the most commonly used material but quartz, sapphire, ferrite and beryllia are used for particular applications. The microstrip is usually thin film because of its well defined line widths, fine lines and high conductivity, but thick films are finding increasing use at the lower frequencies (< 20 GHz). Applications include doppler radar (for traffic control), parametric amplifiers and satellite television. This could turn out to be a massive market if direct broadcasting from satellite becomes commonplace. A hybrid would convert from the 12 GHz broadcast frequency to 400 MHz to enable standard television receivers to be used.

DATA CONVERSION

The feature of hybrids which is most obviously exploited in quality DAC's and ADC's is the versatility to be able to combine several technologies in a single system. Alumina substrates are normally used and these carry a thick-film conductor layer for chip bonding and interconnection. The active devices are custom MOS for low power data manipulation and bipolar for precision amplifiers and reference, and are in chip form because of the large number of interconnections. Resistor arrays are usually thin-film, because of their superior properties, on a small secondary substrate which is bonded to the mother substrate. By using this approach the manufacturer can replace the resistor array if necessary after unsuccessful functional trimming without wasting too much added value.

Figure 9.9 shows a top-end of the market 16-bit DAC by Hybrid Systems which has guaranteed 16-bit monotonicity, can operate with no external components and is TTL and CMOS compatible. Although it contains all the necessary registers, reference and output amplifiers, it dissipates only 450 mW and is mounted in a 24-pin DIP. The very high quality of this sort of hybrid can be judged from some of the specifications :

Linearity	3 ppm FSR/1000 h ;	1 ppm FSR/°C
Gain	10 ppm FSR/1000 h ;	5 ppm FSR/°C
Reference voltage	1 mV/year ;	5 ppm/°C
Offset	100 µV/1000 h ;	3 ppm FSR/°C

This system is representative of the most sophisticated commonly used off-the-shelf hybrids.

The 8-channel, 12-bit data acquisition system shown in figure 9.10 is a very complex hybrid. It contains an 8-channel analogue multiplexer, a sample and hold amplifier, a 12-bit ADC and all the necessary control and interface logic to connect to 8-bit or

Application of Hybrid Technology   153

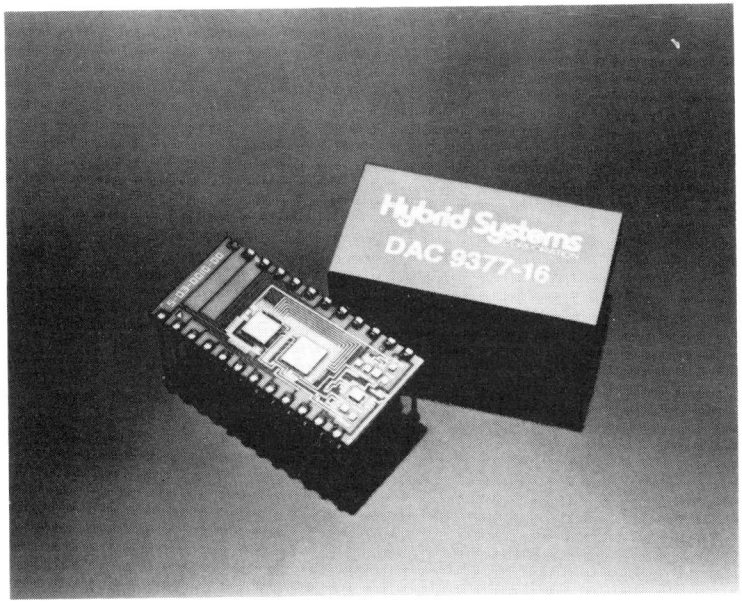

Figure 9.9   A sixteen bit digital to analogue converter
             (courtesy Hybrid (Component) Systems
             (UK) Ltd.)

Figure 9.10 An eight channel, twelve bit data acquisition
             system
                    (courtesy Hybrid (Component) Systems
                    (UK) Ltd.)

16-bit microprocessors; all of this is packaged in a 28-pin DIP. The ADC comprises two chips, one bipolar and one CMOS, to achieve high performance with a low total power dissipation of only 900 mW. The total acquisition and conversion time for this system is only 30 µs. Unusually for such an expensive hybrid, the thin-film resistors are deposited directly onto the substrate. Functional laser trimming is now considered to be reliable enough for the wastage rate of complete hybrids due to overtrimming to be acceptably low.

THE FUTURE

The hybrid market, like any other electronics market, is very volatile - applications are here today and gone tomorrow. Already thin films have almost disappeared except in very precise resistor networks and geometrically precise conductor networks such as microwave devices. In several areas monolithics are taking over from hybrids; this has already happened in some television receivers. ADC's and DAC's are going the same way with precise resistor arrays being replaced by matched current generators, quality 12-bit devices already being available (12). Some active filter applications are likely to be taken over by switched-capacitor filters especially in digital systems where a clock is already available. Clearly semi-custom and custom MOS and ULA technologies will make significant inroads for some other applications.

Nevertheless, there will continue to be a need for a small, reliable, modular interconnection technology and this aspect will dominate over some of the functional features of hybrids in the future. Increasing use of chip carriers will fuel this need as can be seen in figure 9.11 which shows a circuit developed by Plessey Electronic Systems for military applications using PROM's in chip carriers. It is to be hoped that enamelled steel or similar dielectric coated metal substrates will soon be reliable enough to find the place which has been earmarked for them. In some applications they may exploit their special properties such as thermal conductivity and electrically conductive ground plane, but for most applications the important feature will simply be the low cost extension of a convenient interconnection technology. This feature will be augmented if it is coupled with significant

Application of Hybrid Technology 155

Figure 9.11 A military grade hybrid using PROM devices packaged in chip carriers (courtesy Plessy Electronic Systems Ltd.)

developments in the technology of the cheaper polymer thick-film pastes. Lower costs would also seem to lead to the widespread application of copper conductors, but nitrogen firing dielectrics do seem to be difficult to process.

Most indicators suggest that the IC market in Europe will grow at 10% per year, at least to the year 1990, and that the hybrid market will grow at the same rate. Indeed for as long as there is a need to interconnect devices of different technologies in a single system the future of hybrids seems assured. The overriding guideline for system designers who think they have an application which is suitable for a hybrid approach is to take an unblinkered look at the technology and to talk to the hybrid manufacturers at a very early stage.

10
A Customer's View of Hybrid Technology

F. N. SINNADURAI and A. SAUNDERS

ADVANTAGES FROM THE USE OF HYBRIDS

A "customer" may be one who purchases his hybrid microcircuits directly from the manufacturer, or one who influences or requires the use of hybrids in the equipment he purchases. In this context the view contained in this chapter relates primarily to telecomms, but no doubt similar sentiments would apply to many industrial and consumer applications.

Customers' interest in hybrids arose from the need for further miniaturisation of electronic equipment. But rapid progress towards the large scale use of hybrids has been impeded by the strong entrenchment of the dual-in-line package (DIP) and the major investments by equipment manufacturers in automatic insertion machinery. Nevertheless, there remains a strong cost-justification for miniaturisation of equipment because of the high proportion of equipment costs that can be contributed by the PCBs plus their associated connectors, wiring and so on. Depending on the complexity of the equipment, this contribution can exceed 65%, and consequently there is plenty of scope to achieve cost reduction by more efficient use of space on the PCB by miniaturisation of components and higher density interconnection techniques. A move away from the DIP and its interconnection by means of the plated-through-hole (PTH) PCB is certainly justified in any case because of the stagnation of such interconnection costs over the past two decades as shown in figure 10.1 and the continuing rise of PCB material costs in the face of a dramatic fall in IC prices as shown in figure 10.2. Thus the opportunity for reductions in equipment

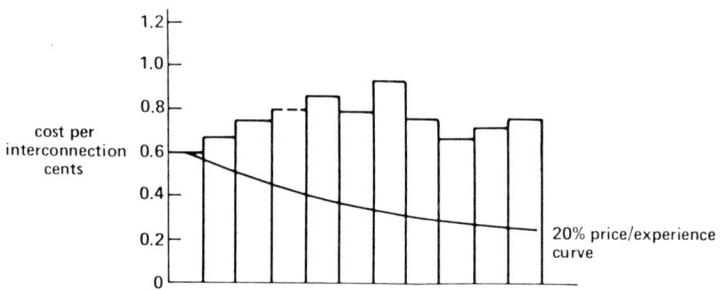

Figure 10.1 Cost trends for interconnection on printed wiring boards.

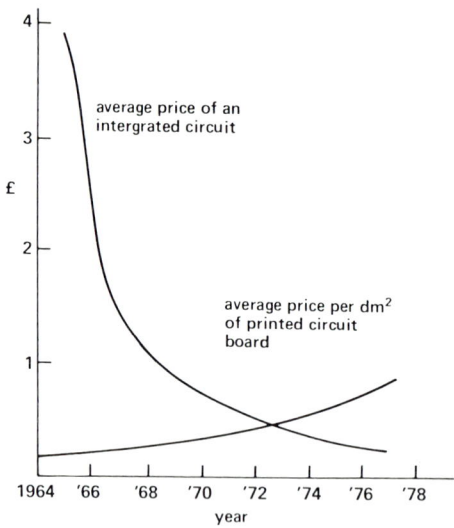

Figure 10.2 Comparison of PCB and integrated circuit prices.

costs by increasing circuit densities through the use of hybrids is there to be grasped.

Potential improvements in reliability, since there are fewer joints, and a reduction in inventory because of the greater extent of integration, are additional advantages offered by hybrid integration. Overall, the eventual advantage is cost, whether achieved immediately through lower prices, smaller inventories, savings in equipment or by lower operating costs throughout the equipment lifetime as a result of the higher reliability.

The technology chosen should be related to the application, and therefore the customer should be certain of his reasons for wishing to use hybrids. The circuit itself - which is often designed by the customer - should then be adapted to suit the technology, because the tolerancing of components from a bread-board design or list of standard components can, for instance, be inappropriate to a thick-film implementation. For example, high-value and tight-tolerance capacitors in RC networks should be designed out, and instead the versatility of thick-film technology employed to achieve a continuous range of resistors that can be made to tight tolerances and will closely track. These resistors may also be trimmed to achieve a given circuit function, hence avoiding trimmer potentiometers. Thus the conventional roles of discrete capacitors and resistors are reversed with hybrid technology, and the wise customer would therefore familiarise himself with the capabilities of the various hybrid technologies.

PERFORMANCE REQUIREMENTS

As stated above, the choices should be related to the performance requirements, and it is useful to examine some examples of choices that have been made in practical applications.

As the various hybrids technologies are described in other chapters, they are not repeated here.

Bare-Chip-and-Wire (BCW) Hybrid Realisations

Compactness in hybrids has often been achieved by directly bonding bare semiconductor chips onto the alumina substrates, and there are indeed requirements that can only be met by BCW technology. For example, in optical fibre systems, the British Telecom approach to realising a receiver has been to use a PIN photodiode detector followed by a GaAs MESFET pre-amplifier. Because the sensitivity of the PIN/FET receiver is inversely proportional to the node capacitance at the photodiode-amplifier interface, it is necessary to achieve a low capacitance at the photodiode and to minimize associated interconnection parasitics. Direct bonding and the elimination of intermediate packaging by the use of BCW techniques, together with small conductor areas and intelligent layout have enabled the parasitics contributed by the hybrid to be substantially lower than the packaged component capacitances and led to the successful realisation of a sensitive detector.

There is a similar sensitivity problem in the optical transmitter modules too. Here, the laser drive and control circuit and the laser chip have been located in close proximity within the same hybrid package in order to minimise the effect of series inductance at high modulation rates and thereby making the layout of the external circuits less critical. Such reductions in path length and size obtained through the use of the BCW hybrid clearly reduce propagation delays and can suit the high bit-rate operations required of optical systems.

Another advantage of BCW technology has been the ability to avoid mismatch caused by the parasitics of an intervening package and enable matched termination to be brought right up to the IC chip. An example is the achievement of the high performance of ECL chips by wire-bonding directly to a 50 ohm impedance thick-film conductor.

The above examples demonstrate some of the advantages that hybrid technology can offer specialist-application high-performance circuits. Indeed the examples cited show a considerable degree of truly "hybrid" integration in which a variety of semiconductor technologies are included. The optical circuits in fact pre-empt monolithic integrated optics by hybrid integration of an optical fibre and, for instance, either a silicon photodiode and GaAs MESFET in the receiver or a GaInAsP laser and dedicated silicon control IC in the transmitter. The freedom to mix a wide variety of technologies and thereby achieve optimum performance by, in this instance, combining GaAs and a mixture of Si technologies such as bipolar, CMOS and NMOS, is thus opened up to the customer.

Hybrid technology has also created the opportunity for the circuit functions to be adjusted after assembly, by trimming thick-film resistors with a laser, thus obviating expensive trimmer potentiometers. A common example of an application is in active filters, where resistors are trimmed to produce the required frequency response.

Miniaturisation has also been a strong motive for using hybrids, and a modest example is the realisation of a synchronising pulse generator which had to fit within the housing of a TV camera for slow-scan TV transmission. This circuit contained 14 integrated circuits. The use of two BCW hybrids employing a simple two-layer thick-film circuit achieved the necessary miniaturisation and occupied an area of about 30 cm^2. Further size reduction was possible by employing multilayer thick-films, but was not justified.

Problems of BCW Hybrid Technology

BCW hybrids can provide the advantages described above, but they can have serious disadvantages too. With the increasing use of ICs in hybrids, manufacturers started to suffer serious problems of poor yield because the bare chips could not be adequately tested beforehand. The graph relating yield to component count is shown in figure 10.3. Some manufacturers have developed their own rigorous electrical probe testing schemes, while others prefer to rework the hybrids. The customer however has a different cause for concern since he has no knowledge of the characteristics of the ICs used in the hybrid, whether they conform at all with his carefully toleranced and specified circuit design and, if not, what margin of safety exists before malfunction occurs. As hybrids become more complex and incorporate more LSI, the ability to characterise adequately the completed hybrid becomes more difficult. Figure 10.4 shows a graph of testability as a function of complexity. Also, the assessment of the composite reliability of the complex hybrid becomes progressively more difficult and less meaningful. Therefore, British Telecom, as an involved customer, has become convinced that there is a strong case to use pre-packaged ICs in the more complex hybrids, which in order to maintain an adequate degree of miniaturisation necessitates the use of micropackages such as Small Outline (SO) packages or Chip Carriers (CCs).

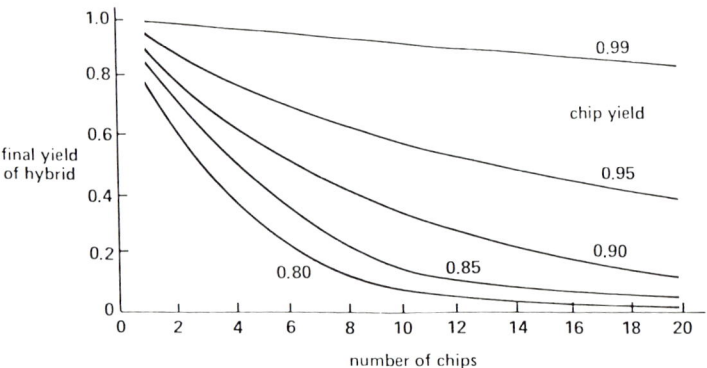

Figure 10.3 Hybrid yield as a function of number of dice and die yield.

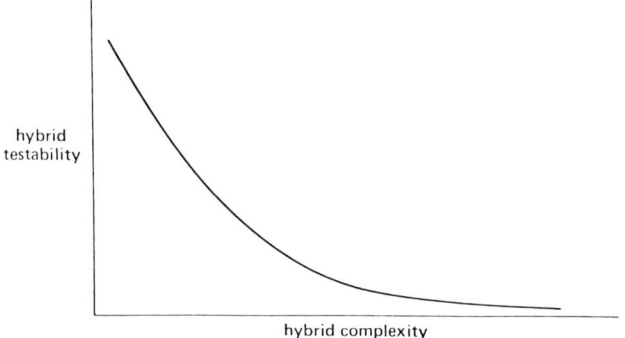

Figure 10.4 The difficulty of testing complex hybrids.

Hybrids Containing ICs in Chip Carriers

This conviction, and the advantage of miniaturisation gained by the use of micropackaged ICs, is illustrated by an example of an Adaptive Equaliser circuit reduced in size for use in a modem. The original circuit had employed TTL ICs in DIPs mounted on a PCB and had occupied 650 cm^2. The use of large-scale-integration (LSI) reduced the ICs to 5 custom LSI chips. But the LSI when packaged in conventional relatively high pincount DIPs and mounted on a conventional PCB with the remaining components, still occupied about 150 cm^2. By micropackaging the LSI components, the two remaining TTL ICs and an EPROM in chip carriers and assembling them on a three-conductor-layer thick-film substrate, it was possible to reduce the circuit to a fully tested hybrid occupying 28.4 cm^2.

RELIABILITY ASSESSMENT

Of course, whilst miniaturisation, performance and immediate cost reductions are important, the customer also has a strong interest in ensuring that the reliability of the hybrids is adequate for his purposes, as the consequences of inadequate reliability would certainly be borne by him. Unfortunately, whilst adherence to functional specifications and quality assurance standards are voluntarily undertaken by manufacturers, the assessment and establishment of component reliability is a less tangible and more onerous burden that has fallen on the customer. Consequently, methods of assessing component reliability have received considerable attention by military, telecomms and computer administrations over the past two decades and, with the support of considerable experimental evidence, have become widely accepted.

Reliable operation means the performance of the required function for the desired duration. The specified longevity for the higher reliability applications is usually about 20 years (175,000 hrs); and the problem that originally faced the reliability engineer was how to simulate such long lifetimes in much shorter timescales that would be acceptable delays when assessing reliability. Thus, time compressions of the order of 200 or more were sought, requiring accelerated ageing tests to be devised to assess component reliability.

Assessment of Semiconductor Components

In order to accelerate the ageing of components, it was first necessary to identify the parameters that contributed to normal ageing mechanisms. The classic relationship that has described reaction kinetics for nearly a century is the Arrhenius equation[1], which relates the rate of reaction (R) and the absolute temperature (T) in the following manner

$$R_T = \text{const} \times \exp(-E_A/kT) \tag{1}$$

Where k is Boltzmann's constant, and E_A is referred to as the activation energy. A graph of this relationship is shown in figure 10.5. Extensive reliability studies have confirmed that the Arrhenius relationship does indeed apply to the thermal ageing processes of semiconductor components, and has led to the concept of accelerated ageing by thermal overstress[2], in which long term operation is simulated by exposing components to elevated temperatures for short durations dependent on the acceleration factor achieved by the overstress.

Attempts were also made to achieve acceleration by voltage overstress, but the scope for increasing voltage above normal operating levels was strictly limited and the technique has not been widely adopted.

As integrated circuits became more complex and were designed with finer line widths, a new failure mechanism, namely electro-migration, became evident as current densities exceeded about 10^9 A m^{-2}. Electromigration[3] proceeds at a rate that varies with the square of the current density (J) as follows:

$$R_J = \text{const } J^2 \exp(-E_A/kT) \tag{2}$$

The result has been confirmed by more recent work[4] and adopted into procurement specifications.

As the use of plastic packaging of ICs became widespread

A Customer's View

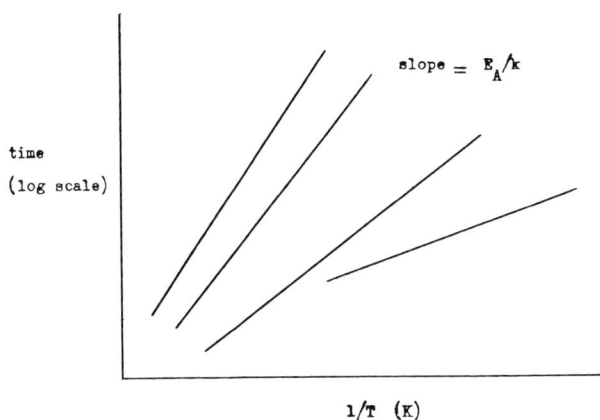

Figure 10.5 Arrhenius plots.

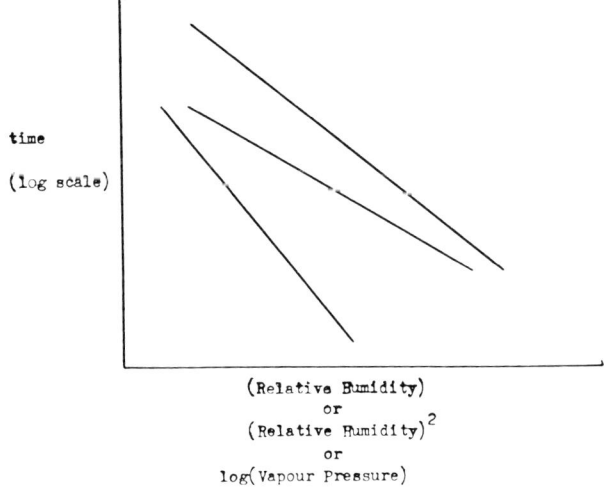

Figure 10.6 Plots of time against moisture (relative humidity) relationships.

and made determined inroads into the former preserves of the "reliable" hermetic packages, a new hazard – arising from the permeation of moisture through the non-hermetic plastic – was recognised, and led to its investigation and the proposal of alternative expressions for accelerated ageing by damp overstress[5,6,7]. Of these, probably the most enduring relationship has been the dependence of ageing rate (R_H) on the square of the relative humidity (RH) as follows:

$$R_H = \text{const} \times \exp(4.4 \times 10^{-4} (RH)^2) \qquad (3)$$

A graph of this relationship is given in figure 10.6. The expressions for acceleration by humidity and temperature may be combined to provide an expression for acceleration (A) by damp heat overstress as follows:

$$A = \exp(0.00044(RH_s^2 - RH_{amb}^2) + E_A/k (1/T_{amb} - 1/T_s)) \qquad (4)$$

In which the suffixes "s" and "amb" refer to the stress and ambient conditions respectively. Unfortunately the values of E_A/k applied in calculating tests for reliability approvals are often very low and overcautiously pessimistic, and result in excessively severe reliability tests.

The above expressions apply to integrated circuits, whether they be independently packaged or integrated within a hybrid microcircuit, and the factor that dominates will depend very much on whether it is humidity or temperature that prevails in the operating environment.

Assessment of Thick-Films

With the advent of hybrid microelectronics, the new elements that were introduced were the film circuits; and new attention was necessary to establish methods of accelerating ageing of these elements. Studies of thick-film resistors[8,9] confirmed that they too obeyed the Arrhenius relationship, but with a different activation energy from that for the semiconductor components.

A Customer's View

Acceleration by humidity was also obtained, but with an RH index of 1 and not 2 as obtained with ICs. Hence, the combined expression for acceleration by damp heat is as follows:

$$A = \exp(0.025(RH_S - RH_{amb}) + 8120(1/T_{amb} - 1/T_S)) \qquad (5)$$

It is interesting to consider some of the observations of the ageing behaviour of commercial thick-film resistors:

The most consistent behaviour exhibited by thick film resistors has been a progressive increase in value that varies approximately with the square root of time, and the changes increase with temperature and humidity. These relationships are shown in figure 10.7 and 10.8. Extrapolation of these changes back to the temperatures and humidities of normal operation, have revealed that most commercially-obtainable thick-film resistors would drift by less than 0.5% in 20 years, and some would undergo less than 0.1% drift[9], an excellent prospect indeed for thick-film resistors for high reliability applications. Some anomalous behaviour has also been observed, in which plastic encapsulation of iridium oxide based resistors caused the drift to increase dramatically from less than 0.5% unencapsulated to more than 5% when encapsulated. Anomalous sensitivity to the magnitude and polarity of applied voltage has also been observed, whereby negatively-biased iridium-oxide resistors showed large positive drift, and the drift could be slowed or reversed by reversing the applied bias.

It is clearly essential, in order to safeguard the customers' interests, that both normal ageing and anomalous behaviour should be detected during reliability assessments. Fortunately, as both were found to be accelerated by thermal overstress and damp-heat stress, it is possible to apply routine standard tests for reliability assessment, provided that the conditions that would provoke the anomalous behaviour - such as encapsulation in plastic, and electrical biasing - were applied to the thick film resistors, and provided of course that they are relevant to the application. Such accelerated ageing tests have been designed and incorporated, for instance, into the British Telecom Generic (Procurement) Specification D4500[10] for Hybrid Microcircuits.

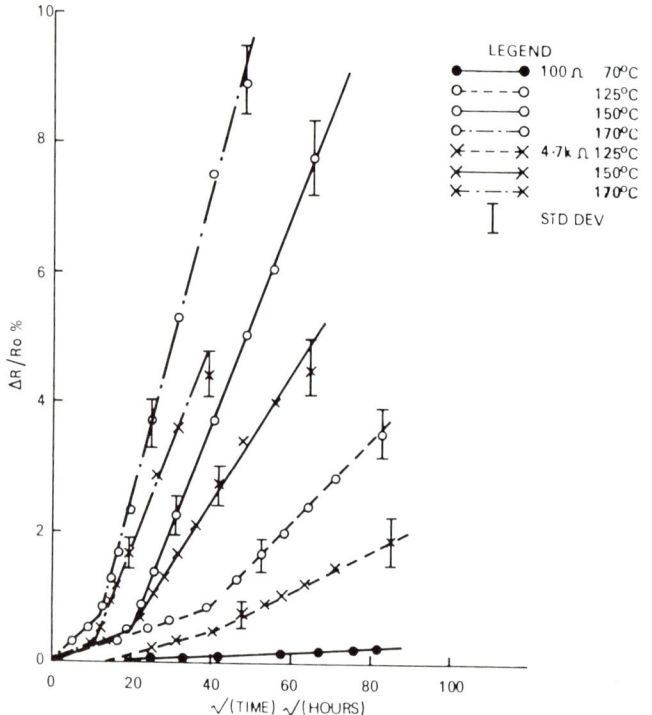

Figure 10.7 Variation of resistance as a function of time at four different temperatures for one particular manufacturer's product.

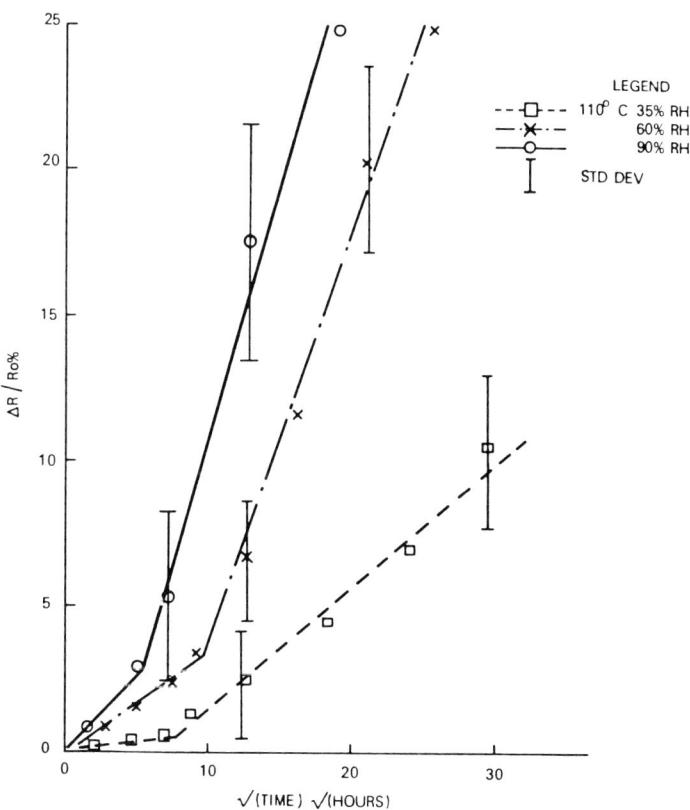

Figure 10.8 Variation of resistance as a function of time at three different relative humidities for one particular manufacturer's product.

Whilst such assessments can help avoid the procurement of unstable components, ideally it would be best if the thick-film circuits are actually manufactured in a manner suited to achieving high reliability. The customer, who bears the consequences of unreliability, has the strongest motive to ensure that the technologies employed are suited to his reliability requirements. (In other words he has the strongest motive to find and specify the solutions to the problem he would otherwise be saddled with).

Examples of procedures that are expected to achieve higher reliabilities from thick-film resistors and conductors include: overglazing with a compatible glaze; avoiding plastic encapsulation; locating sensitive resistor elements away from high voltage elements; ensuring symmetry of geometry and electrical layout of resistors required to track closely; designing the size and trim of close tolerance resistors to limit power dissipations to less than 16 mW/mm^2 in the residual width of the resistor.

Assessment of Active Hybrids

The hybrid microcircuit of course incorporates added active components such as diodes, transistors and integrated circuits. If the semiconductor components were pre-packaged, then their reliability assessment would be in accordance with the accelerated ageing tests described earlier. However, the demand for high packing densities led designers to opt to use the semiconductor devices in bare-chip form, wire-bonded directly to the hybrid substrate. Because hermeticity has been regarded as essential to achieve high reliability, new (and large) hermetic packages were developed to be able to contain the hybrid substrate. With few expensive exceptions, the bare chips as well as the ceramic substrates were glued down (and indeed still are) with organic adhesives within the hermetic package. These new technologies were adopted and taken for granted with very little understanding or evidence of reliability performance. The new environment within the encapsulation was in fact likely to be different from that encountered by the individual ICs and thick-film networks. Nevertheless, provided that no new failure mechanisms were induced in either the ICs or the thick-films, then a conventional thermal overstress could be regarded as a relevant reliability assessment test. However, reliability studies of the BCW technology - undertaken once more by a customer[11] - revealed that a new hazard was indeed introduced by the new technology: namely the organic adhesives outgassed ammonia and/or water vapour. The results of one particular test are shown in figure 10.9. As the vapours were trapped within the hermetic enclosure, they were concentrated around the very components that were supposed to be protected by the package. Thermal overstress did reveal very poor reliability indeed due to the use of amine cured adhesives as shown in figure 10.10, but the elimination of ammonia did not achieve a great improvement because alternative adhesives involved more water vapour instead. A correlation between outgassing and weight loss from the adhesive was also established and is given in figure 10.11. This provides a means of assessing the adhesives beforehand[12]. Elimination of the substrate adhesive completely, by integrating the substrate with the package base in the "Integral Substrate Package" (ISP), did achieve a marked improvement in reliability and these results are also shown in figure 10.10.

Use of the ISP did not incur a cost penalty for the reliability advantage gained, and indeed actually created opportunities for further cost reductions by the use of low cost materials. Although the ISP provided a probable solution to the problem of the unreliability of the hermetic hybrid package, the use of bare-chip ICs gave rise to the other serious problems of characterisation and reliability assessment that were described earlier.

THE DEVELOPMENT AND EVALUATION OF RELIABLE PLASTIC MICROPACKAGING

Despite the arguments already advanced for micropackaging, it appeared that the high cost and lack of international standards for chip carriers - which were initially available only in hermetic form - still remained a disincentive, and that users and manufacturers were each waiting for the other to take the initiative. Steps towards standardisation of CCs and SO packages were taken by JEDEC in 1981 and 1982, but the cost disadvantages remained. As progress towards miniaturisation, and consequent advantages to the customers, was clearly not proceeding at an adequate pace, new initiatives to overcome the impasse had to be taken.

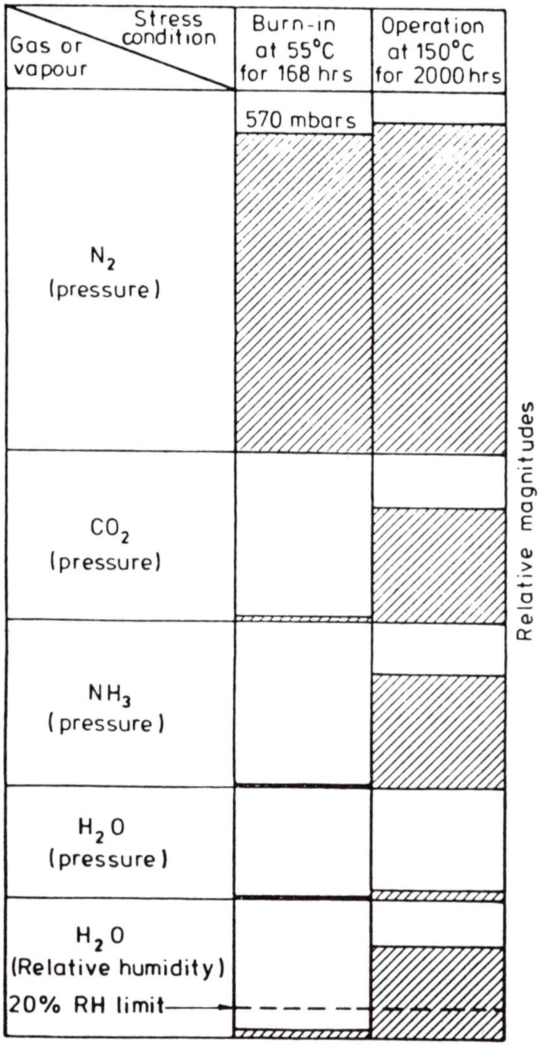

Gas or vapour	Stress condition	Burn-in at 55°C for 168 hrs	Operation at 150°C for 2000 hrs
N_2 (pressure)		570 mbars	
CO_2 (pressure)			
NH_3 (pressure)			
H_2O (pressure)			
H_2O (Relative humidity) 20% RH limit			

Figure 10.9   Analysis of the gas contained within a hermetic hybrid enclosure before and after life test.

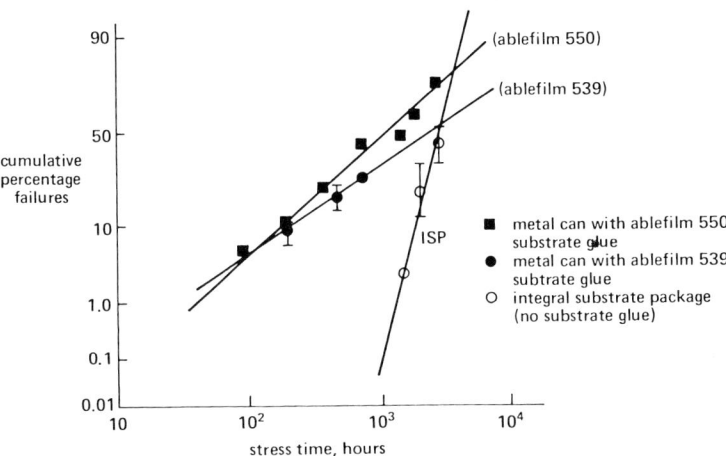

Figure 10.10 Comparison of cumulative failure distributions of various hermetic Telecoms hybrids overstressed at 150°C.

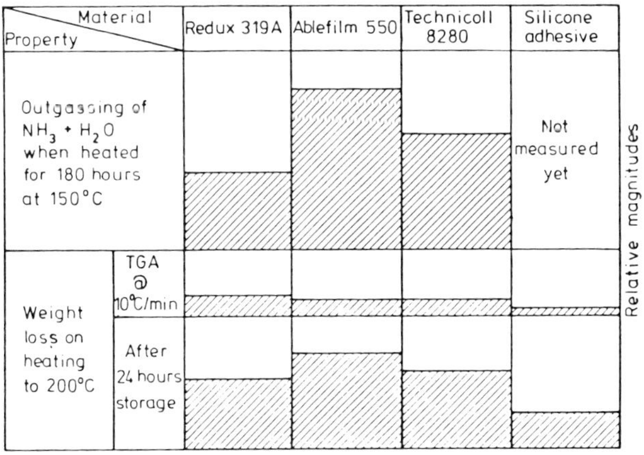

Figure 10.11 Correlation between outgassing and weight loss for various adhesive used in hybrid technology.

As a customer with a strong interest in gaining the expected benefits from low-cost micropacking, BT has made extensive studies of plastic coatings with the aim of achieving low-cost but reliable encapsulation of integrated circuits. Evaluation of a number of junction coatings (i.e. those used in direct contact with the IC chip) and combined junction-plus-top coatings (used to provide mechanical protection) were conducted by employing severe and extended overstress tests[6]. These materials are given in figures 10.12 and 10.13 respectively. Especially high acceleration of ageing by damp-heat stress has been made possible by using a non-saturating autoclave technique originally devised by British Telecom[13], in which the typical stress condition is 108°C, 90% RH. A set of accelerated ageing factors due to this stress relative to some possible operating ambients, have been calculated from equation (4) employing a low activation that convincingly safeguards the customers interests and the resulting figures are given in table 10.1.

Clearly quite large acceleration factors are obtained for ageing in temperate climates, and very short duration tests - of up to 100 hours - would simulate 20 years operation. Longer testing of up to 2000 hours is necessary to simulate continuous tropical operation at 35°C, 90% RH. However 35°C, 90% RH is unlikely to occur continuously over 20 years, even in the tropics, and a more realistic simulation would probably be around 500 hours at 108°C, 90% RH. Nevertheless, some really excellent results, as shown in figure 10.14, were obtained from two particular combinations of coatings which successfully protected the encapsulated components even for 2000 hours (i.e. the full tropical stress condition!). Other coatings performed less effectively and the majority were totally inadequate[6]. A table of merit, given in figure 10.15, shows the preferred coatings obtained from this particular exercise, which may be identified by reference to figures 10.12 and 10.13.

Such findings clearly showed that high reliabilities were attainable with low cost plastic coatings and, as an involved customer, British Telecom pursued this line of development and devised a new low-cost plastic chip carrier to be used with the plastic coatings. Manufacturers also committed themselves to production of ICs in SO plastic micropackages, the most notable

Coating Code	Coating Type	Curing Temp °C
A	Rigid silicone	175
B	Semi-flexible silicone	175
C	Flowing silicone	175
D	Thixotropic silicone	175
E	Silicone-epoxy specially adapted	160
F	Poly-paraxylene	-
G	Heat-resistant polyimide	175
H	Unfilled epoxy (junction or top coat material)	160
J	Filled epoxy	160
L	Filled silicone (junction or top coat material)	175
M	Thixotropic epoxy dip (top coating material)	120
D+H	Silicone + Epoxy	160
D+M	Silicone + Epoxy	120
R	Proprietary combination: junction + top coatings	-
S	Proprietary combination: junction + top coating	-

Figure 10.12   Details of junction coatings used.

Coating Code	Coating Types	Final Curing Temp °C
C+L	Thin silicone + Filled silicone	175
C+M	Thin silicone + Thixotropic epoxy	120
C+N	Thin silicone + Fusing epoxy powder	100
C+P	Thin silicone + Phenolic dip	160
D+L	Thixotropic silicone + Filled Silicone	175
D+M	Thixotropic silicone + Thixotropic epoxy	120
D+N	Thixotropic silicone + Fusing epoxy powder	100
D+P	Thixotropic silicone + Phenolic dip	160
J+L	Filled epoxy + Filled silicone	160
J+M	Filled epoxy + Thixotropic epoxy	120
J+N	Filled epoxy + Fusing epoxy powder	100
J+P	Filled epoxy + Phenolic dip	160
K*+L	Silicone RTV + Filled silicone	175
K*+M	Silicone RTV + Thixotropic epoxy	120
K*+N	Silicone RTV + Fusing epoxy powder	100
K+P	Silicone RTV + Phenolic dip	160

Figure 10.13   Details of junction plus top coatings evaluated.

TABLE 10.1

ACCELERATION FACTORS AND TEST DURATION AT $108°C$, 90% RH
RELATIVE TO SOME HUMID OPERATING AMBIENT CONDITIONS

	Temperature °C	Relative Humidity %	Acceleration	Test Duration hrs.
UK Telephone Exchange	30	25	3100	60
UK Office	20	45	3700	50
UK Uncontrolled	12	72	1800	100
Tropic Uncontrolled	35	90	90	2000

A Customer's View 177

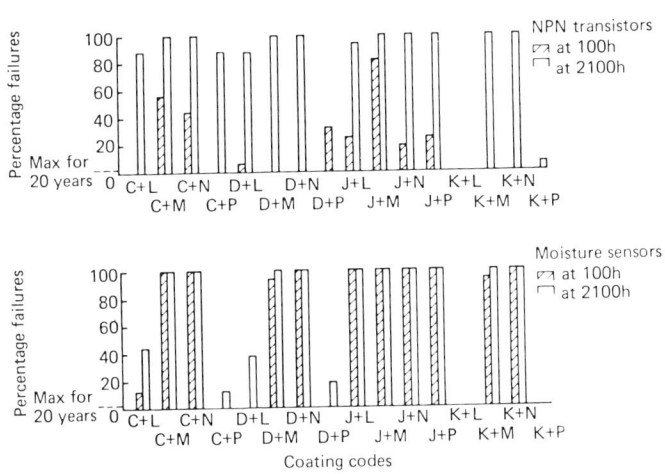

Figure 10.14 Bar charts of percentage failures of test vehicles for the combinations given in figure 10.13.

Order of Merit	Junction Coat + Top Coat
1	K + P
2	K + L
3	C + P
4	C + L
5	D + P
6	D + L
7	J + P
8	J + L
9	J + N
10	K + N

Figure 10.15 Order of Merit of coatings tested.

being Philips, followed by SGS, Ferranti, and more recently, a number of others. Regrettably the long-promised post-moulded plastic chip-carrier has not yet appeared in production volumes. The reliability prospects of ICs plastic-coated in the BT chip carrier (made by PCB techniques, hence "PCB" CC) have been compared with similar ICs in SO micropackages in extensive reliability evaluations[14] which revealed that most types would achieve 20 year lifetimes, as shown in figures 10.16 and 10.17, with the PCB CC encapsulation producing the outstanding result. These results are summarised in figure 10.18 in an order of merit for package type and IC chip passivation.*

Such results have significantly advanced the prospects of early cost reductions of micropackages, and should assist in breaching the "DIP DAM" (i.e. the present entrenchment of the DIP). The forecasts, given in figure 10.20, are that once the dam is breached, DIPs will be substituted at an increasing pace by micropackages[15]. And the corresponding trend in costs, as shown in figure 10.19, predict that if the traditional requirement for hermeticity prevails, then interconnection costs would fall below the DIP and associated PTH mounting by the end of the decade; but if the reliable plastic micropackages are adopted instead, then the cost advantage should be realised by the middle of the decade.

As other chapters have described, the hybrid also incorporates passive add-on components and these too have their interconnection cost trends. Together, these cost trends, the earlier justification of miniaturisation, the case for using micropackaged rather than bare-chip ICs, and the reliabilities that are

---

* Editor's Footnote

This new and promising technique is protected by, inter alia, European Patent Application 82300596.2 filed by British Telecom with F.N. Sinnadurai, A.J. Cook and K.W. Gurnett as the named inventors. It should be noted that all the reliability studies reported so far have been with the PCB CC mounted on similar PCB material. At the time of writing, the thermal performance of the PCB CC has not been reported.

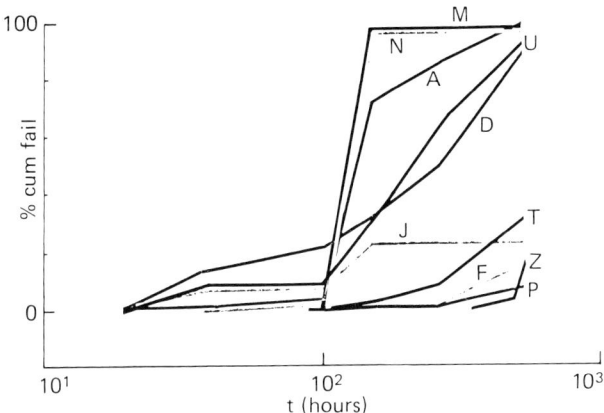

Figure 10.16 Cumulative failures of 741 and 348 amplifies from various sources in SO, LCC and DIL packages overstressed at 108°C, 90% RH.

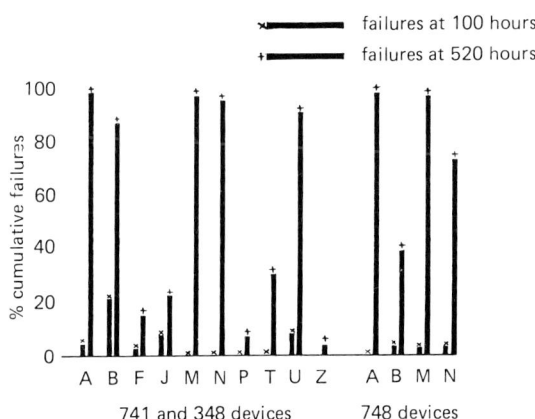

Figure 10.17 Percentage cumulative failure of operational amplifiers from various sources overstressed at 108°C, 90% RH.

1. Ti-Pt-Au metallised, nitride passivated ICs from source B assembled in junction coated and lidded PCB CC.
2. Al metallised, oxide passivated ICs in moulded plastic SO8 packages from source E.
3. Ti-Pt-Au metallised, nitride passivated ICs from source B assembled and junction coated on single layer ceramic CC.
4. Al metallised, oxide passivated ICs from source B assembled and junction coated on single layer ceramic CC.
5. Al metallised, oxide passivated ICs from source V assembled and junction coated on ceramic CC.
6. Ti-Pt-Au metallised, nitride passivated ICs in moulded plastic SO8 packages from source B.
7. Al metallised, oxide passivated ICs in moulded plastic SO8 packages from source B.

Figure 10.18 Summary of order of merit for various encapsulations and IC manufacturing sources.

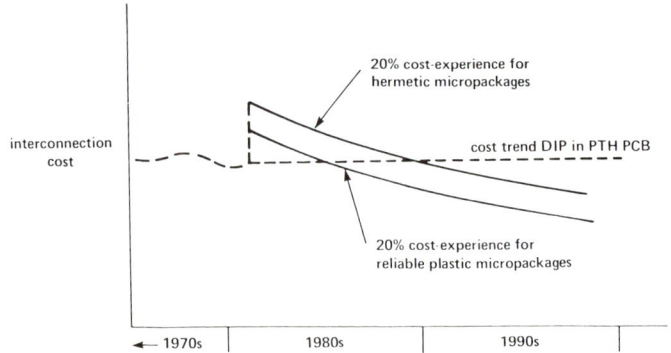

Figure 10.19 Predicted cost-per-interconnect trends for DIL packages inserted in plated-through-hole PCB and surface mounted micropackages.

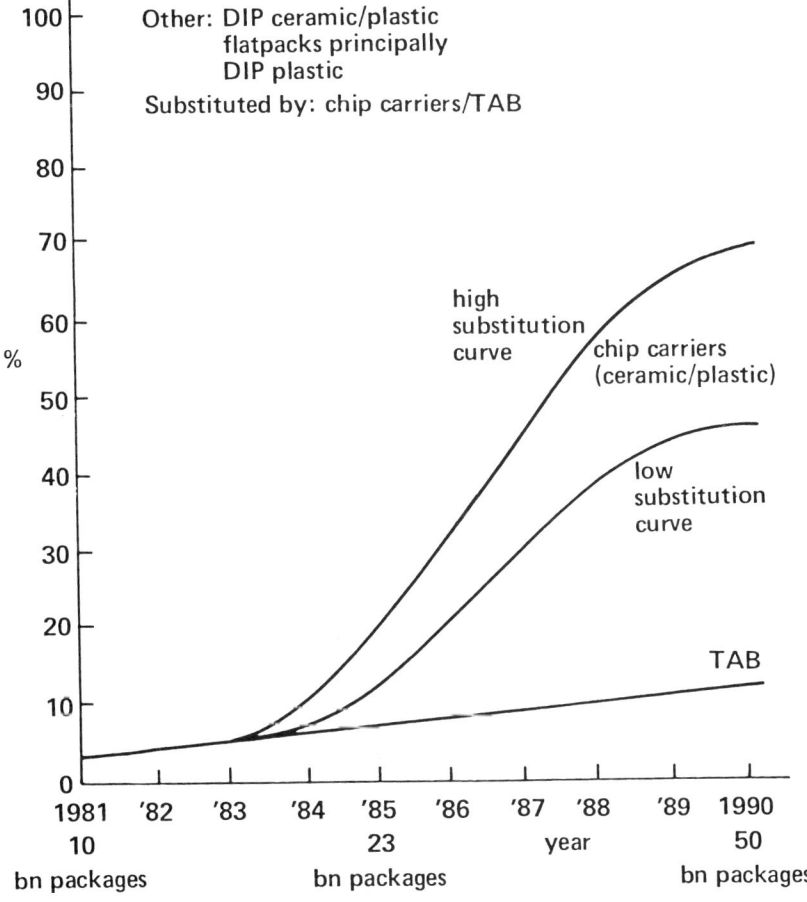

Figure 10.20 Predicted substitution curves for the replacement of DIL packages by chip carrier and TAB techniques.

demonstrably attainable by thick-films and plastic encapsulated ICs, show that there are distinct advantages to be gained from the use of active and passive microcomponents and thick-film elements to achieve high density "hybrid" integrated circuits. In this context the "hybrid" will increasingly not be restricted to a component made with alumina substrates and mounted on a circuit board, but will eventually replace entire circuit boards. Such a trend will involve a significant shift in value-added assembly from the equipment manufacturer to the hybrid manufacturer (who may well become an in-house operation under the control of the equipment manufacturer), and some upheaval will undoubtedly occur in the presently established organisation of manufacturing. Hopefully, this progress will not be impeded by those with vested interests in old technology. Thus the benefits described above should be realised right through to the ultimate customers.

11
Quality Control and Assurance in Hybrid Circuit Production

J. T. LYNCH

INTRODUCTION

One of the prime advantages of using thick film hybridisation for the production of microcircuits lies in the flexibility of the technique both with regard to layout and materials as well as packaging parameters. This can, however, create severe disadvantages if the build standard and procurement of a hybrid are not equated with its fitness for purpose. Disparities can, at one end of the spectrum, result in hybrids required for high reliability applications failing during service or, at the other extreme, commercial hybrids being too expensive because of over-specification. Most unsuitable hybrids result from a lack of quality assurance in the design stage coupled with poor liaison between the customer and hybrid manufacturer. In many cases a hybrid has been designed into a circuit because of a single advantage that a hybrid module could bestow without due consideration of several possible accompanying disadvantages. Often, the hybrid manufacturer is unaware of the functions of the subsystem or total equipment and produces a module fulfilling the requirements of the procurement specification and related test conditions which does not work in the field. Instances also exist where orders have been placed upon and accepted by hybrid manufacturers whose technologies and capabilities are not compatible with the build requirements. This has inevitably produced costly delays, sometimes halting the progress of complete equipments and has unnecessarily resulted in damaged relationships and loss of confidence in the hybrid industry as a whole.

PROBLEMS ASSOCIATED WITH HYBRID PROCUREMENT - SIX CASE HISTORIES

Most UK hybrid manufacturers currently possess some form of Approval to BS9450. Nevertheless, by virtue of the fact that thick film hybrids are generally custom-built and sometimes cannot be purchased to a National Approval, problems in the procurement of hybrids still do exist. A number of studies* of thick film hybrid circuits has shown that many of the problems could have been avoided with the correct use of National Standards and customer quality assurance. In order to illustrate the type and variety of problems that can occur a selection of the summaries of the studies is presented here. As the use of the Standards is increasing, examples of such problems are becoming rarer and most of these case histories are several years old. The importance of Customer Quality Assurance was initially discussed some time ago (1) and its role is now fully recognised. The emphasis here is on the steps the hybrid manufacturer can take to prevent the types of problems illustrated from occuring in the first place.

Case 1: Hybrid Failure - No Liaison in Design, Manufacture or Test

The thick film hybrid module was required in a high reliability category for an aerospace application but no malfunction warning or backup redundancy was designed around the component into the system. A detailed failure analysis was performed after a hybrid had failed in the field. The following were noted :-

(a) The modules were encapsulated in epoxy and mutual thermal mismatch between the substrate, the epoxy and large add-on components had caused substrate damage and open circuits between substrate and add-on component. This is shown in figure 11.1.

---

* These case studies were carried out at the Quality Assurance division of Plessey Research Ltd. This is an independent division and carries out assessment and failure analysis work for a variety of organisations. The examples presented are from several different thick film manufacturers.

(b) Neither the epoxy used for encapsulation nor the silver-palladium conductor ink were suitable for an aerospace application.
(c) Injudicious cropping of redundant leads and irregular lead forming of a flat pack together with incorrect track pitching due to poor layout design had resulted in substandard reflow solder assemblies as well as possible damage to the glass wall of the flat pack.
(d) Hand soldering coupled with multiple heat applications in additional jointing in a localised area had resulted in conductor leaching and the overall standard of soldering was unsatisfactory.
(e) The modules were not capable of passing temperature cycling or vibration tests implied by the application. These, however, were not detailed in the purchase specification.

It was concluded that the design and build standard were totally unsuitable for an aerospace application, one of the major causes being the subcontracting of the module manufacture through three companies without adequate liaison. The hybrid manufacturer was unaware of the final use of the hybrid or the performance requirements and had not performed vital environmental testing or employed the correct methods of manufacture.

Case 2: Hybrid Yield Problem - No Design Analysis or Worst Case Analysis

The hybrid in question contained a number of integrated circuits of the same design obtained from one manufacturer. These were standard chips specified by the customer and purchased to the I. C. manufacturer's specification. Such specifications allow variations in some of the electrical parameters. Too great a mismatch of one particular parameter within the module resulted in rejection of the module on a functional electrical test but this had been allowed for in the manufacturer's costing and the overall hybrid yield was acceptable. The variation of this parameter, did however, result in unacceptable performance when several such hybrids were placed in a sub-system together.

This problem, which caused costly delays, could have been

Figure 11.1 An example of open circuits caused by the thermal mismatch between the substrate, component and epoxy encapsulant.

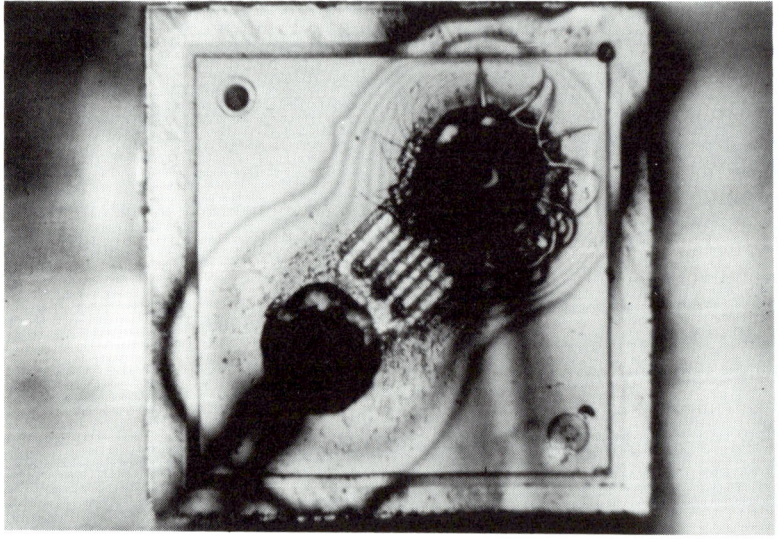

Figure 11.2 Example of transistor damage caused by excessive power during bonding.

avoided by a suitable worst case tolerancing analysis in the design stage and a more complete liaison between the customer and hybrid manufacturer.

Case 3: Hybrid with Unique Add-on Component (Single Sourced)

The use of a custom-built add-on component in a custom-built hybrid is always inadvisable especially if the add-on is of complex design. Investigation of failures of a transistor array designed into a thick film hybrid concluded that the failures resulted from an extremely poor build standard coupled with inadequate inspection on the part of the component manufacturer. It transpired that the arrays had been assembled in a development environment with minimal quality control rather than on a production line. The manufacturer was unwilling to start up a production line with proper quality controls for the initial volume requirements and second sourcing proved impossible. Approximately £2,500 was expended in a failure analysis, respecification and, ultimately, redesign. The problems associated with the procurement of this one component contributed greatly to the loss of an equipment contract. Examples of defective workmanship are shown in figures 11.2, 11.3 and 11.4.

Case 4: Hybrid with Frozen Design

The hybrid manufacturer's Rule Book defines the limits of the capabilities of his technologies. There are instances where the manufacturer may exceed these limits under controlled conditions but total disregard of the Rule Book can produce disastrous results. A manufacturer places himself in such a predicament when, for one reason or another, he accepts a frozen design from a customer which demands technologies he does not currently operate. Such shortcomings were observed in a series of hybrids. The dimensions of the hybrids and their package terminations were pre-determined by the customer's mother-board. These constraints and the environmental requirements necessitated a package and a sealing technique outside of the manufacturer's technology as did the customer specification on chip attachment. The resultant modules were seen to be of experimental rather than production standard and subjected to excessive rework. The assessment report concluded that no confidence could be placed in any hybrid produced by the

Figure 11.3 Example of a poor standard of wedge bonding.

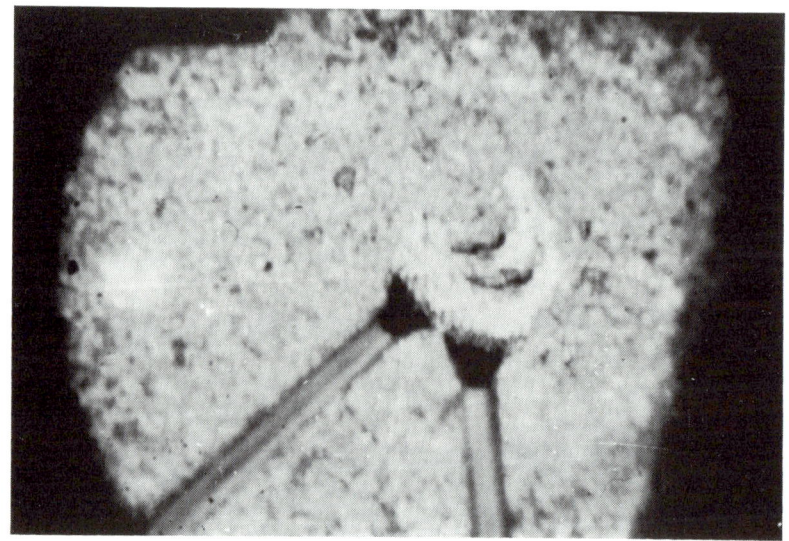

Figure 11.4 Two wedge bonds superimposed.

manufacturer to this particular design. Figure 11.5 shows the effect of chlorine contamination on an aluminium wirebond.

Case 5: Hi-Rel Hybrid Produced by Consumer-Grade Manufacturer

Failure analysis of a hybrid designed for long term reliability equipment resulted in a vendor assessment of the manufacturer's facilities. The following deficiencies were amongst those noted :-

No written instructions for manufacture or inspection
No retention of production or inspection results
No formal route cards
No regular equipment calibration
No control of drawings or documents
No goods inwards inspection
No segregation of rejected items

The vendor's problems were compounded by an unreasonable buying policy in as much as a printed circuit board realisation would have been superior to the specified hybrid which was :-

Not a functional entity
Could not be tested
The price demanded by the purchaser was too low

This case history illustrated the necessity for adequate vendor assessment prior to the placing of an order. It also produces an example of a manufacturer prepared to attempt a high reliability product without the necessary facilities.

Case 6: Resistor Network Sub-contracted for Package Sealing

Samples received from the customer for detailed failure analysis showed open-circuit failure modes, some intermittent. After decapping, each substrate was found to have been coated with a thick layer of junction coating resin which is normally used in thin layers for protecting semiconductors. This had been used by the manufacturer in an attempt to protect the resistor network whilst the lids and packages were furnaced for solder sealing by an outside contractor. The subcontracting was forced upon the manufacturer by his own lack of facilities. As a result of the

Figure 11.5 An example of the effect of chlorine contamination on an aluminium wire-bond.

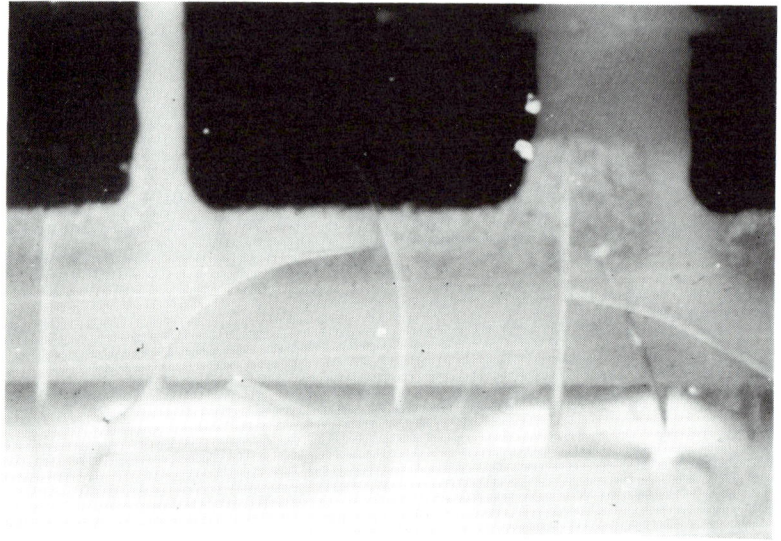

Figure 11.6 Wire bond fractures caused by cracking in the encapsulant.

furnace treatment extensive cracking occurred in the resin causing the severance of wire bonds and the consequent open-circuit failure mode. The failure of these thick film resistor networks was compounded by the very real possibility that units already installed in boards would also develop the faults as cracks propagated. Figure 11.6 shows wire bonds severed by cracks in the resin.

By buying down to a price, following inadequate design assurance and vendor survey, the customer was faced with the prospect of replacing units already installed in equipment. This resulted in costly delays to a multi-million pound military project.

PURCHASING TO NATIONAL STANDARDS

The BS9000 System and BS9450

U. K. industry has taken a significant lead in standardising the design, manufacture and procurement of hybrids whilst still allowing the technique its full flexibility. Most manufacturers are approved to supply hybrids under the British Standards BS9000 system. The portion of BS9000 relevant to thick film hybrids is BS9450 which is a Capability Approval system. Unlike Qualification Approval where the build standard is fixed to a single device type, Capability Approval allows a manufacturer to produce an infinite range of fully approved hybrids provided that materials, techniques, dimensions, performance and testing lie between declared limits.

The way in which BS9450 operates is to assess the capability of the manufacturer and his process to manufacture the particular device required. The manufacturer is called upon to produce a "Capability Manual" which defines his process, methods of assembly, design rules, tolerances and limiting values of his process. The manufacturer is also required to produce and test a number of "Capability Qualifying Components" (CQC) which embody the boundary, or worst case conditions of the claimed capability work for which he is seeking BS9450 Capability Approval. The CQC's must be subject to a stringent approval programme before approval is given to BS9450.

Continuous monitoring of the process is then carried out by testing the CQC's periodically. The tests to be applied to the CQC's are given in BS9450 and consist of the specific "Group A" and "Group B" tests which must be carried out on a lot-by-lot basis (but where a lot can represent up to one month's production), "Group C" tests which are carried out every 3 or 6 months and "Group D" tests which are carried out every 1, 2 or 3 years.

The customer's hybrid is detailed in the Customer Detail Specification which is raised by the customer but with assistance from the manufacturer and to guidelines laid down in BS9450. This gives the customer further confidence that the hybrid is being procured to a high standard of specification and test. The tests specified in Subgroup AO of Group A (Electrical Tests) are carried out on each of the customer's circuits.

The tests specified in the remainder of Group A are carried out with each production batch of the customers circuit forming the inspection lot (i.e. on a statistical sample) and the Group B tests are also carried out on a representative inspection lot (including CQC's). The tests of Group C and Group D are carried out on the CQC's covering the design and performance boundary conditions employed in the customers circuit but, by arrangement, any of these tests could be done on the customer's hybrid.

TESTING TO BS9450 - THE INSPECTION REQUIREMENTS

The Origin of the Inspection Requirements

The test procures detailed in the inspection requirements of BS9450 have been derived from BS9300 and BS9400 which were originally written as British Standards for discrete transistors and integrated circuits. Although a thick film hybrid can contain many diverse add-on components and be considerably bulkier than those devices envisaged by the first issues of BS9300 and BS9400, the test procedures have needed little modification to encompass the changes in technology dictated by hybrids.

The origins of the BS9000 series of inspection requirement test procedures can be traced back to the basic reasons for testing a

product. Probably the most fundamental reason is to reveal
weaknesses, potential or otherwise, before the customer finds out
the hard way. The BS9000 systems divides inspection requirements
into Groups A, B, C and D.

Group A contains non-destructive, short duration electrical
and mechanical (although in practice, mainly electrical) tests which
are employed to assess on a lot-by-lot basis the principal
characteristics of the component.

The tests in Group B of BS9000 are designated as electrical,
mechanical or environmental and are carried out lot-by-lot. These
are all intended to be carried out in less than a week and in some
cases may be destructive. Apart from the Electrical Endurance
Test the BS9000 Group B tests were originally drawn-up to simulate
the variety of events which might befall an electronic component
between manufacture and commission. Thus, there were tests
relating to handling, storage at extremes of temperature and
humidity and solderability. Some of the mechanical tests, however,
are more commonly performed during the periodic testing of
Groups C or D in most current BS9000 Specifications.

The Group C tests of BS9000 are designed to cover electrical,
mechanical, environmental and endurance characteristics of a
component which are appropriate for inspection at regular intervals
of three or six months. In other words, these tests are done
periodically, rather than lot-by-lot, to verify characteristics
additional to the more important ones of Groups A and B.

The tests in Group D are designated in BS9000 as being
normally carried out at intervals of one, two or three years. The
tests in Group D are considered as a periodic check on declared
characteristics (including those necessitating tests of long durations)
and also a means of collecting information required for presentation
in the certified test records. In some ways the Group D tests are
more of a check that there have been no design or process changes
rather than a periodic check on quality.

In Groups B, C and D of BS9000 the component is subjected
to Electrical Endurance testing. In Group B the test lasts for
160 hours and can be conducted under one of a number of electrical

test conditions. The test is conducted at the maximum possible temperature without exceeding the thermal rating of the device. The value of this test and the extension of it in Groups C (2000 hours) and D (8000 hours) need to be examined in the light of its derivations from BS9300 and BS9400. For 8000 hours endurance testing of a TTL device at 125°C, it can be roughly equated to 50 years life under normal operating conditions at 70°C. If failure rate with time of such a discrete device is assumed to obey the classical bath tub curve, successful completion of the 8000 hour test is an indication that the end of life wear-out phase has not yet begun. Similarly accelerated operation of the discrete device for 160 hours simulates avoidance of early failure (infant mortalities) whereas 2000 hours is an assessment of the medium term life or flat portion of the bath tub curve. Thus, with a discrete device it is possible to gain a degree of insight into the short, medium and long term life expectancy. With a thick film hybrid, the complexity and variety of components make any such calculation much more difficult but, nevertheless, the three electrical endurance tests of Groups B, C and D provide some confidence as to reliability of the manufacturer's hybrids. Further information on accelerated aging can be found in reference 2.

Group A Testing

There are two Subgroups to Group A, viz. A1 and A2. Subgroup A1 comprises a simple external visual examination to ensure legibility and correctness of markings and lack of damage to the encapsulation and leads. Subgroup A2 is used to assess the major static or dynamic characteristics of the hybrid at 25°C and is basically utilised to verify the description or characteristics of the component.

Group B Testing

The Group B tests in BS9450 commence with Subgroup B1 which is subdivided into B1(a), Solderability and B1(b), Dimensions. The first assesses whether the leads of the device can be readily soldered and the second whether the device has the correct dimensions. Those dimensions measured at this Subgroup are generally only the critical interchangeability parameters.

Subgroup B2 is designed to stress the device mechanically by temperature cycling (generally between the storage temperature limits) followed by firstly an electrical check and then either a leak test or exposure to damp heat cycling. The philosophy of the B2 test sequence is that the required minimum of 5 complete temperature cycles, whilst not representing the device life, will turn a latent defect into an obvious failure. In other words, the test duration is not designed to show up a mediocre device. The test is intended to stress die bonds, wire bonds, package seals and, where applicable, the integrity of leads through package walls. Any damage to the circuit within the package is then obvious when the device is electrically tested. Where the temperature cycling has resulted in damage to the package hermeticity, the fault can be recognised for a cavity package by leak detection or, in the case of non-cavity packages (e.g. plastic encapsulation), by damp heat cycling. The leak test is performed in two steps called fine and gross. The commonest form of fine leak test consists of placing the device in a chamber which is evacuated, back filled with a specified pressure of helium and left for a period of time. It is assumed that for a given pressure and time the helium will pass through any holes or cracks into the main cavity of the package at a rate proportional to the size of the hole or crack. On removal from the helium "bomb" the helium flows back into the atmosphere and the rate at which it does so is again proportional to the size of the hole or crack. If this flow rate is between certain limits the helium can be detected by the mass spectrometer of a commercial leak detector. This particular test is sensitive to leak rates between 10^{-8} and 10^{-5} atm. cm^3/s (10^{-3} and $1Pa\ cm^3/s$). It is generally assumed that a device with a leak rate of better than 10^{-8} atm. cm^3/s is "hermetic" but theoretical calculations show that within a short period of time compared with anticipated device lifetime the associated leak path has allowed a complete change of atmosphere. How relevant this is to real life is not clear! Where the leakage path is very large the helium forced into the package flows out very quickly and by the time the device is in the mass spectrometer it may well have disappeared, giving the appearance of a pass. It is, therefore, essential to complement fine leak testing with a gross leak test. In essence this is done by placing the device in a hot liquid which causes the gas in the package to expand and bubble out of the hole.

The damp heat cycling test, in conjunction with a subsequent electrical test, is designed to detect flaws in non-cavity encapsulations. At Group B the requirement is for 6 cycles which is, like the Group B temperature cycling, designed to detect faulty devices, whereas the 28 day test in Group C gives a degree of confidence that the device will last for an acceptable life in a panclimatic environment (other than shipborne above deck). The actual conditions of the test are approximately 95% RH and cycling between $25°C$ and $40°C$ or $55°C$. Meteorological measurements made over many years have shown that a relative humidity of $>95\%$ combined with a temperature $>30°C$ does not occur in free air conditions over long conditions except in regions of extreme climate. However, these conditions can occur in confined spaces such as vehicles, tents, and aircraft cockpits. Thus, although the test has an accelerated element it is not entirely divorced from a possible worst case situation.

Subgroup B3 of BS9450 calls for a check on terminal robustness and offers a choice of test methods dependent upon the actual termination. The test is designed to demonstrate that the device is capable of withstanding the potential rough handling which may result when, for example, the units are inserted into boards.

The acceleration test called up in BS9000 is intended to assess the satisfactory performance of components when subjected to forces produced by steady acceleration environments (other than gravity) such as occur in moving vehicles, especially flying vehicles, rotating parts and projectiles. The standard levels of test quoted in BS2011 range from 6g to 30,000g which are put into context when it is realised that the maximum acceleration achieved by a space rocket is only in the order of 500g (although this can produce associated ringing and flexing transients up to about an order higher). Nevertheless, the acceleration test in Subgroup B4 of BS9450 still offers a range of standard severities between 10g and 30,000g, not to simulate life but to apply an artificial stress on wire bonds, die bonds, and other attachments. If a stress level of 20,000g is considered on a wire bond the associated force is somewhere in the order of 1 gram whereas in the case of a large semiconductor chip the force is about $\frac{1}{2}$ kilogram for the same acceleration. If applied to a large (say 64 lead) chip carrier, the force rises to about 40 kilograms. For the wire bond and the chip the

forces involved are really only testing for the poorest quality of
bond and thus provide a useful back-up check once the lid has been
sealed on. The situation is different for large components such as
the chip carrier, since although the solder joints should comfortably
bear the 20,000g and its associated 40kg force, the acceleration
could cause flexing, bowing and even cracking of the mounting
board which would result in premature joint fracture. It is,
therefore, essential for both manufacturer and customer to set
a realistic value of acceleration.

Subgroup B5 subjects the device to 160 hours of Electrical
Endurance and is the first of three such tests (as discussed above).
Subgroup B6 of BS9450 designates the post test end points (i.e.
that is to be carried out at the end of subgroup B2, B4, B5) in order
to ensure that the devices are still working. In the case of BS9450
it is a repeat of Subgroup A2, the major static/dynamic
characteristics at 25°C.

Subgroup B7 is entitled CTR (Certified Test Results)
Information and requires that the manufacturer lists the results of
the tests carried out at Subgroups B1, B2, B4 and B5. These results
must be authenticated and copies given to BSI and the NSI and the
manufacturer is obliged to make available, on request, copies of
current CTR's to bona fide users. CTR's thus provide a potential
customer with a degree of quality and reliability information on
the manufacturer's products.

Group C Testing

Subgroup C1 is sub-divided into two sections, the first, C1(a),
is designed to place two different kinds of mechanical stresses on
the devices followed by damp heat exposure to probe any damage
caused by the previous mechanical tests. Vibration is the first
of the two mechanical tests and most manufacturers offer Vibration
Swept Frequency, where subsamples are vibrated in one of three
mutually perpendicular axes for two hours, one subsample per axis.
For small components such as most of those covered by BS9450, a
resonance search is not considered necessary and is not called up by
the test. The severity of the test can be chosen from a list given
in BS9450. The frequency range 150-2000 Hz is generally
considered to be representative of frequencies found in a real life

environment (such as rail transport, tanks, aircraft, etc.) and, except for large packages, is offered by most manufacturers. Again, except for large packages, most manufacturers offer the maximum severity of vibration amplitude which is associated with a corresponding value of acceleration amplitude. This is 1.5 mm and 20g (196 m/s^2). The sweep rate is set at approximately one octave per minute which, for the full duration of the test, gives something in the order of one million reversals. Vibration is followed, using the same samples, by shock, a test which is designed to simulate the effects of relatively infrequent, non-repetitive shocks likely to be encountered by equipments in service or during transportation. In other words, the test is not designed to prove components which are required to exist in an environment of repeated shocks and jolts. The shock tester is, in essence, a simple machine. The sample is securely fixed to a table which can be allowed to fall under gravity along rods or runners. On the underside of the table is a hammer which, at the end of the fall, impacts on to a bed of material (or anvil) with absorbant properties (usually rubber, lead or a metallic honeycomb). The combination of table height above the bed, hammer geometry and properties of the bed material define the shock experienced by the component on the table. The BS9450 requirement is for the shock to have a half sine pulse shape with a choice of severities. Most manufacturers offer a peak acceleration of 14,700 m/s^2 with a 0.5 ms pulse or 981 m/s^2 with a 6 ms pulse which is monitored by attaching a sensor coupled to a storage oscilloscope to the shock tester. Having completed both mechanical tests the samples are subjected to damp heat testing which can be cyclic as in Subgroup B2 or Steady State. In the former case, the duration of the test is set at 28 cycles, rather than the six cycles of Subgroup B2 and is designed not only to pinpoint damage which has occurred to the packaging by the mechanical testing but also to give a limited degree of confidence that an undamaged package will provide a lifetime of protection.

Subgroup C1(b) is a periodic assessment of the device dimensions and is intended to check the less important parameters not covered by Subgroup B1(b).

Subgroup C2 is also divided into two sections, both designed to cover electrical measurements not performed during the lot-by-lot testing of Subgroup A2. Subgroup C2(a) is a repeat of the Subgroup A2 test but performed at the maximum and minimum temperatures of operation. Subgroup C2(b) is an assessment at ambient temperature of those parameters not measured at Subgroup A2 (i.e. minor static or dynamic characteristics).

Subgroup C3 is the extension of Electrical Endurance started during the Subgroup B5 test and continues from 160 hours up to (generally) 2000 hours on a smaller sample representative of production throughout the period.

As in Group B, Group C ends with post test end points for Subgroups C1 and C3 (Subgroup C4) and the CTR requirements for Subgroups C1 and C3 (Subgroup C5).

Group D Testing

Subgroup D1 is divided into two sections, D1(a) being a repeat of Subgroup C1(a) but only being applicable for Basic Approval, whereas in C1(a) the tests are only applicable for Full Approval. Subgroup D1(b) is the check on those dimensions not covered by Subgroups B1(b) and C1(b).

Subgroup D2 contains three tests which are applicable to plastic encapsulated devices only and they are, in effect, a check on the plastic material and the application. Subgroup D2(a) is Resistance to Solvents Test and is basically a measure of how well the plastic encapsulation (solid or cavity) can stand up to various levels of exposure to a variety of solvents. Subgroup D2(b) is applicable to non-cavity plastic packages and is a check on the flammability of the plastic. The test can be performed either by passing an excessive current through the packaged device or by holding it in a standard flame. The final test, Subgroup D2(c) is Rapid Change of Temperature, involving placing the plastic encapsulated devices alternately into liquids at $-40°C$ and $+100°C$ through ten complete cycles to ensure no mechanical or electrical damage. The three sections of Subgroup D2 require different samples since each test is destructive.

Resistance to solder heat, Subgroup D3 is a periodic test to ensure that the device can withstand the application of solder heat

(350°C) for 3 seconds. It is designed to make allowances for variations in the solder heat that a device might see during installation.

Subgroup D4 is the final leg of the Electrical Endurance Test and takes the samples up to 8000 hours.

The Inspection Requirements are completed by the Post Test End Point Measurements for Subgroups D2(c) and D4 (Subgroup D5) and the CTR information requirements for Subgroups D2, D3 and D4 (Subgroup D6).

AQL's, Inspection Levels and Sampling Plans

The number of hybrids examined during the lot-by-lot or periodic testing and the number of failures permissible without condemnation of the lot is defined by the stated AQL and Inspection Level of the Sampling Plan. The definitions of these terms and the operation of the sampling plan are explained in BS6001, "Sampling Procedures and Tables for Inspection by Attributes". Acceptable Quality Level (AQL) is defined as the maximum percent defective (or the maximum number of defects per hundred units) that, for the purposes of the sampling inspection, can be considered satisfactory as a process average. In other words, if a consumer designates an AQL of 4% for a certain defect, he is indicating to the supplier that his (the consumer's) acceptance sampling plan will accept the great majority of the lots or batches that the supplier submits, provided that the process level of defectives is no greater than 4%. Nevertheless, this does not give the supplier the right knowingly to supply any defective units.

The sampling plan itself indicates the number of units from each lot or batch which are to be inspected and the criteria for determining the acceptability of the lot or batch. The Inspection Level determines the relationship between the lot or batch size and the sample size. The Inspection Level will obviously be dependent upon how critical the parameter to be inspected is considered. The various categories, Level I, II or III, S-1, S-2, S-3, S-4 are defined in BS6001 together with the complete Single Sampling Plan for Normal Inspection and various other sampling plans. The mathematics of the various sampling plans are also explained in

BS6001 using Operating Characteristic Curves which indicate the percentage of lots or batches which may be expected to be accepted under the various sampling plans for a given process quality. As an example, if a manufacturer produces 10,000 units, of which a randomly distributed 100 (1%) are defective, then a random sample of 80 could pick out anything between 0 and 80 defectives. If the customer requires 80 to be inspected (Inspection Level I) and sets an AQL of 1%, the lot is accepted (according to the sampling plan for Single Sampling, Normal Inspection) with 2 or less rejects out of the 80. Examination of the appropriate operating characteristic curve shows that for every lot of 10,000 examined according to this plan there is approximately a 95% probability that the lot will be accepted and a 5% probability that the sample will contain more than 2 defectives (out of the total 100 available) and therefore that the lot will be rejected. The original figures for AQL's in BS9300 and BS9400 were arrived at, not by design requirements or even a wet finger, but by analysis of manufacturers' test results. In general the process average for performance was about an order better than the designated AQL for that particular property or parameter. This was done in order to ensure that the manufacturer would expect to pass the test on every occasion if he maintained that level of quality. The Inspection Levels were laid down in order to allow the minimum sample from the operating characteristic curve. In other words, instead of needing to choose say, 200 samples and being allowed 10 failures, it was set so that 8 could be selected with 1 failure allowed.

PURCHASING TO OTHER NATIONAL STANDARDS

Procurement of hybrids within the general pattern of BS9450 may follow a number of paths. The hybrid can be purchased to BS9450 directly, to other related national specifications such as the British Telecomms. documents RC5394 and D4500 or to a Company's own specification. RC5394 and D4500 require that the hybrid manufacturer shall release to BS9450 Schedule A together with further stipulations on materials and testing.

DEVELOPMENT OF CUSTOMER STANDARDS

Even procurement of a thick film hybrid to BS9450 does not guarantee that the hybrid will be suited for its purpose since, in general, the hybrid is custom-built. In some cases there is insufficient interchange of information between customer and manufacturer and sometimes it may well be that for security reasons the customer is unable to divulge information on the usage and environmental requirements of the hybrid. It is not entirely unknown for a customer to refuse to give the manufacturer the performance specification of the hybrid!

Because of these and other limitations it is essential that all personnel concerned with the hybrid's design, procurement and evaluation on the customer's side are aware of the problems of the technique as applied to their individual customer function.

One solution to the manifold problems that are encountered by a prospective user/purchaser of hybrid circuits is the provision of some form of guide or standard to the use and procurement of hybrid circuits. The very fact that hybrids are custom-built, and therefore, require a positive input from the customer means that every company, no matter how large or how small, needs some written guidelines on the subject. The very large company may require a Standard for each division of the company if those divisions have diverse requirements.

The overall purpose of any document should be to tell the customer what a hybrid is, why he might want to use one and how he would go about procuring one which is suited for his purposes. Descriptive passages on hybrids, the various styles, packaging variants, add-on components, and how the hybrid is made, can take many forms all of which should be tailored to the needs of those using the guide. One universal requirement of any guidelines is an indication of the advantages of hybrids compared with other technologies (such as printed circuit board assemblies or silicon integrated circuits) which might perform the same electrical function. Similarly, once a customer has chosen the hybrid route, every guide should contain a section on how to go about the procurement, from choice of suppliers and initial negotiations, through prototyping and on to the goods inwards procedures for production quantities.

A short description of an existing standard may help other companies who wish to follow a similar path. The Standard was prepared for the use of those design engineers concerned with the introduction of thick film hybrid circuits into equipment. As well as indicating the required sequence of events from hybrid conception to module production the Standard provides ancillary information with precautionary advice.

The introductory section is followed by sections comparing the relative merits of :

Thick film hybrids
Thin film hybrids
Silicon integrated circuits
Printed circuit board technologies

Both practical and technological comparisons are made which, initially at least, should allow the user to decide whether a thick film hybrid is indeed the best choice. This decided, it then becomes essential to ensure that the hybrid consists of a functional unit which is capable of being specified and tested in isolation. Rulings and advice on these considerations are included and guidelines on performance specifications, choice of component values and tolerances and statistical computer-aided analysis are laid down.

A section on hybrid components gives some educational detail on substrates, the various inks, solder add-on discretes, chip and wire bonding and packaging. The main purpose of the chapter is, however, critical comparisons of, for example, the various types of :-

Conductor inks
Discrete component attachment
Chip and wire attachment methods
Packaging techniques

These comparisons are made with regard to the compatibility of the parameters one to another as well as to the ultimate requirements of the hybrids. Essential layout rules universal to the hybrid manufacturing industry are also included.

Further sections emphasise the necessity for the early and

disciplined liaison with prospective suppliers, including :-

   Dual sourcing arrangements
   Technical, environmental and performance considerations
   Prototypes
   Production timescales
   Information flow
   Tooling agreements

Case histories of hybrid failures such as those related above have resulted in a section highlighting the various problems and pitfalls and suggestions for their avoidance. Further emphasis is laid on multi-sourcing and early liaison between customer and vendor; the concept of the design cycle is explained and the major design considerations are presented. These include :-

   Spatial requirements
   Thermal requirements
   Tolerancing
   Printed component requirements
   Suggestions for the usage of discrete components
   Packaging parameters

Categorisation of hybrids has been undertaken as part of a design policy in order to aid in the production of blank procurement specifications as well as limiting the choice of hybrid suppliers for the production of any one hybrid. It was decided that the categorisation should comprise of a two dimensional matrix whose variants were style complexity and packaging. This resulted in fifteen generic categories of hybrids. An ongoing programme of vendor assessments has resulted in a list of approved suppliers for each generic category and this will be updated periodically. Because of the importance attached to procurement specifications for thick film hybrids blank procurement specifications have been produced for each of the generic categories. Advice is given on the filling in of a blank specification for any hybrid within a particular category and guidelines are presented on the inspection and test requirements as a function of the manufacturer's BS9450 Approval and Capability Qualifying Circuits.

The Guide is completed by a number of appendices including

a step by step description of activities necessary to achieve satisfactory procurement of an acceptable thick film hybrid. The information is also presented in a flow diagram, indicating backward movement upon the failure of any activity.

ACKNOWLEDGEMENTS

The support and collaboration of Plessey Telecommunications and Office Systems Ltd., is acknowledged.

References and Bibliography

CHAPTER 1

General References

1. HARPER, C.A.: <u>Handbook of Thick Film Hybrid Micro-electronics</u>, McGraw Hill, (1974)

2. HOLMES, P.J. and LOASBY, R.G.: <u>Handbook of Thick Film Technology</u>, Electrochemical Publications Ltd., (1976)

CHAPTER 2

1. PULFRICH, H.: Ceramic to Metal Seal, US Patent 2163407, (June 20, 1939)

2. PULFRICH, H.: Vacuum Tight Seal, US Patent 2163408, (June 20, 1939)

3. MAISSEL, L.I. and GLANG, R.: <u>Handbook of Thin Film Technology</u>, McGraw Hill, (1970)

4. Permanent Interconnection Technology, ERC Working Party Report, (January, 1980)

5. DRYDEN, W.G.: Design Guidelines for Thick Film Hybrid Circuits, Electronic Packaging and Production, pp 140-145, (July, 1979)

6. HOLMES, P.J. and LOASBY, R.G.: Handbook of Thick Film Technology, Electrochemical Publications Ltd., (1976)

CHAPTER 3

1. Electronic printing compositions and vehicles therefor, British Patent 1,518,926, E I Du Pont de Nemours, (July, 1978)

2. Improvements in or relating to vitrifiable fluxes, British Patent 803,943, E I Du Pont de Nemours & Co., (5th November, 1958)

3. LOASBY, R.G., DAVEY, N. and BARLOW, H.: Enhanced property thick-film conductor pastes, Solid State Technology 15, No. 5, pp 46–50, (May, 1972)

4. SMITH, B.R. and DIETZ, R.L.: An Innovation in Gold Paste, Proceedings ISHM 1972 International Microelectronic Symposium, pp 2-A-5-1 2-A-5-8, (November, 1972)

5. ANGUS, H.C. and GAINSBURY, P.E.: Glaze resistors with ruthenium dioxide, Electronic Components, (January, 1968)

6. Oxides of cubic crystal structure containing bismuth and at least one of ruthenium and iridium, US Patent 3,583,931, E I Du Pont de Nemours, (June, 1971)

7. WALTON, B.: Principles of thick film materials formulation, Radio and Electronic Engineer 45, No. 3, (March, 1975)

8. LAURIE, A.S.: A high quality, base metal, thick film resistor system, Proceedings IEEE 23rd Electronic Components Conference, pp 137–139, (May, 1973)

9. Crystallizable glass composition, British Patent 1,182,987, E I Du Pont de Nemours & Co., (4th March, 1970)

10. LOASBY, R.G.: Aspects of multilayered thick-film hybrids, Solid State Technology 14, No. 5, pp 33–37, 46, (May, 1971)

11. DELANEY, R.A. and KAISER, H.D.: Multiple-curie-point

capacitor dielectrics, I.B.M. Systems 11, No. 5, pp 511-9, (September, 1967)

12. BOWKLEY, I.C.: Improved, glass-ceramic, thick film capacitors, Proceedings Conference on Hybrid Microelectronics, Canterbury, (IERE Conference Proceeding No. 27), pp 47-55, (September, 1973)

13. COLEMAN, M.V. and GARNETT, G.E.: Surface area, structure, and composition of debased alumina substrate, Proc. IERE Conf. on Hybrid Microelectronics, Loughborough, (September, 1975)

14. SCHABACKER, R.B.: Porcelain enamelled substrates for hybrid circuits and printed circuits, Proc. European Conf. Hybrid Microelectronics, Ghent, (May, 1979)

15. VERMEIRCH, G. et al: Comparative study of thick film Cu systems for multilayer hybrids, 4th European Hybrid Microelectronics Conf. Copenhagen, (May, 1983)

16. STEIN, S.J., HUANG, C.Y.D. and KELLY, J.M.: Polymer Thick Film Materials, Proc. 3rd European Hybrid Mircoelectronics Conference, Avignon, (May, 1981)

CHAPTER 4

General References

1. TOPFER, M.L.: Thick Film Microelectronics, Van Nostrand Reinhold Co., (1971)

2. HAMER, D.J. and BIGGERS, J.V.: Thick Film Hybrid Microcircuit Technology, John Wiley & Sons Inc., (1972)

3. MILLER, L.F.: Thick Films Technology and Chip Joining, Gordon and Breach, New York, (1972)

4. RIKOSKI, R.A.: Hybrid Microelectronic Circuits - The Thick Film, Wiley - Interscience, (1973)

CHAPTER 5

1. Uncommitted substrate material of the type described is available from :

    Balzers Akteingesellschaft,
    FL - 9496 BALZERS,
    Principality of Leichtenstein.

2. EADES, J.D., JORDAN, R.G., KELLY, R.G. and MURRAY, J.: Application of GAELIC to the Design of a Large Scale Integrated Circuit, International Conference on Computer Aided Design, IEE Conference Publication No. 111, (April, 1974)

3. DOIG, R.C.: Computer Aided Design of Thin Film Circuits, International Conference on Computer Aided Design, IEE Conference Publication, No. 175, (July, 1979)

4. NC controlled flat-bed drawing machines and tape controlled rubylith cutting knives are available from :

    Ferranti: CETEC Graphics,
    Bell Square,
    Brucefield,
    Livingston,
    West Lothian.

5. LAW, H.T.: The Role of Thin Hybrid Microcircuits in Electronics, Third International Conference on Thin Films, "Basic Problems, Applications and Trends", Budapest, Hungary, (1975), and Thin Solid Films 36, 323-329, Elsevier Sequoia S.A., Lausanne, (1976)

CHAPTER 6

General References

1. PLANER, G.V. and PHILLIPS, L.S.: Thick Film Circuits, Butterworth, London, (1974)

2. JONES, R.D.: Hybrid Circuit Design and Manufacture, Marcel Dekker Inc., New York, (1982)

3. Hybrid Microcircuit Design Guide, published by ISHM, Montgomery, Alabama, (1982)

CHAPTER 7

General References

1. CORKHILL, J.K.: Packaging of Hybrid Microcircuits, Chapter 2, from HOLMES, P.J. and LOASBY, R.G.: Handbook of Thick Film Technology, Electrochemical Publications Ltd., (1976)

2. BURGGRAF, P.S.: Semiconductor Package Sealing, Semiconductor International, (September, 1979)

3. SWAVING, E.C.J. and GUILONARD, P.J.G.: The metallization of different aluminas by the molybdenum-manganese process, from The Use of Ceramics in Valves, published by The British Ceramic Research Association.

4. JONES, G. and WATERFIELD, B.C.: Design Considerations for Ceramic Packages, International Packaging and Production Conference, Internepcon, Brighton, (October, 1972)

5. ERIKSON, G.: Chip Carriers - Coming Force in Packaging Electronic Packaging and Production, Vol. 21, No. 3, pp 64-80, (1981)

6. BOETTI, A., LYNCH, J.T., McCARTHY, J.P. and HEPHER, M.R.: Investigation into microjoining techniques for "hi-rel" active chips, Proc. European Hybrid Microelectronics Conf. Avignon, France, pp 75-88, (1981)

CHAPTER 9

1. ZECHNALL, W.: Hybrid microelectronics in modern car radios, radios, ISHM 80, New York, pp 245-248, (1980)

References

2. ANG, L.Y., and LEADBETTER, L.D.: Hybrid circuit for anti-lock braking application in heavy truck environment, ISHM 78, Minneapolis, pp 276-279, (1978)

3. DELL'ACQUA, R. and FORLANI, F.: Progress in hybrid technology in Italy, ISHM 79, pp 247-254, Los Angeles, (1979)

4. ARIMA, H., IKEGAMI, A., ABE, K., IWANAGA, S. and ISOGAI, T., Thick-film sensor, ISHM 80, New York, pp 272-277, (1980)

5. IWANAGA, S. and IKEGAMI, A., Thick-film humidity sensor, IEEE Conf. Proc., New York, pp 58-66, (1981)

6. DELL'ACQUA, R., DELL'ORTO, G. and VICINI, P.: Thick-film pressure sensors: performances and practical applications, Third European Hybrid Microelectronics Conference, pp 121-Avignon, pp 121-134, (1981)

7. ANDERSON, C.M. and NAGUIB, H.M.: The Development of 8" - 100 lpi thick-film thermal printheads, ISHM 80, New York, pp 201-208, (1980)

8. LIN, H.S.: The Hybrid microelectronic technology impacts on the implantable pacemaker, ISHM 76, Vancouver, pp 188-191, (1976)

9. AUCOUTURIER, J.L., CANIVENC, R., MARQUES, M. and GOVAERTS, R.: Biotelemetry and radiotracking of wild birds: portable device using solar cells power supply, ISHM 77, pp 1-7, (1977)

10. DAY, S.M.D. and PATTERSON, F.K.: What's new in European thick film?, New Electronics, pp 54-58, (5th May, 1981)

11. BOKIL, D. and MORONG, W.: Thick-film transformer advances hybrid isolation amplifier, Electronics pp 113-117, (25th August, 1981)

12. EMMENS, T.: Data conversion - the monolithic future, New Electronics, pp 36-38, (2nd November, 1982)

13. BARNWELL, P.: A novel, economic, high density hybrid assembly, ISHM 82, pp 155-157, Reno, (1982)

CHAPTER 10

1. ARRHENIUS: Z Physical Chem, Vol. 4, pp 226, (1889)

2. REYNOLDS, F.H.: Thermally Accelerated Ageing of Semiconductor Components, Proc IEEE, Vol. 62, No. 2, pp 212-222, (February, 1974)

3. BLACK, J.R.: Electromigration - A Brief Survey and Some Recent Results, IEEE Trans Electron Devices, Vol ED16, pp 338-347, (1969)

4. SIM, S.P.: Procurement Specification Requirements for Protection Against Electromigration Failures in Aluminium Metallisations, Microelectronics and Reliability, Vol. 19, No. 3, pp 207-218, (1979)

5. PECK, D.S. and ZIERDT, C.H.: Temperature-Humidity Acceleration of Metal-Electrolysis Failure in Semiconductor Devices, Proc. 11th Annual Reliability Physics Symposium, pp 116, (1973)

6. SINNADURAI, N.: An Evaluation of Plastic Coatings for High Reliability Microelectronics Journal, Vol. 12, No. 6, pp 30-38, (1981)

7. LAWSON, R.W.: The Accelerated Testing of Plastic Encapsulated Semiconductor Components, Proc. 12th Annual Reliability Physics Symposium of the IEEE, pp 103-112, (1979)

8. SINNADURAI, N., SPENCER, P.E. and WILSON, K.J.: Some Observations on the Accelerated Ageing of Thick Film Resistors, Proc. European ISHM Conference, Ghent, pp 113-121, (1979)

9. SINNADURAI, N. and WILSON, K.J.: The Ageing Behaviour of Commercial Thick-film Resistors, IEEE Trans on Components, Hybrids and Manufacturing Technology,

Vol. CHMT-5, No. 3, (September, 1982)

10. Generic Specification for Thick-film Microcircuits, D4500, Issue C, British Telecom Systems Evolution and Standards Department SES4.1.3, (1983)

11. SINNADURAI, N., WILSON, K.J. and BRACE, D.W.J.: Some Problems and Possible Solutions for Hybrid Microcircuit Reliability, Microelectronics Journal, Vol. 11, No. 1, pp 26-36, (1980)

12. British Telecom Specification M219F for "Type Approval and Quality Assurance of Encapsulating Resins, Coatings and Adhesive Materials for Semiconductor Devices and Microcircuits", BT Quality Assurance Dept., MSP7/QA5, London Materials Science Section.

13. SINNADURAI, N.: The Accelerated Ageing of Plastic Encapsulated Semiconductor Devices in Environments Containing a High Vapour Pressure of Water, Microelectronics and Reliability, Vol. 13, No. 1, pp 23-27, (February, 1974)

14. SINNADURAI, N. and ROBERTS, D.: Assessment of Micropackaged Integrated Circuits for Hybrids in High Reliability Applications, Proc. International Symposium on Hybrid Microelectronics, Reno, (15-17th November, 1982)

15. BPA Technology and Management, Report on "Integrated Circuit Packaging and Equipment Interconnect Practice", (1982)

CHAPTER 11

1. LYNCH, J.T.: The Role of Customer Q.A. in reducing Hybrid Reliability Problems, Microelectronics and Reliability, Vol. 16, pp 523-526, Pergamon Press (1977)

2. REYNOLDS, F.H.: Thermally Accelerated Ageing of Semiconductor Components, Proc. IEEE, pp 212-221, (February, 1974)

General References

3. BS9450 : 1975 (1980), Specification for custom-built integrated circuits of assessed quality : Generic Data and Methods of Test.

4. NF C 96-410 (December, 1982), Circuits Integres a Couches et Hybrides.

5. Generic Specification for Film and Hybrid Integrated Circuits (Draft CECC Specification).

6. MIL STD 1772 Certification Requirements for Hybrid Microcircuits Facilities and Lines.

About the Authors

NORMAN G. BURROW was awarded his BSc degree in Electrical Engineering from the University of London, King's College in 1967. He pursued a programme of research in the same college on the subject of positive metal ion condensation on clean metal surfaces and was awarded his PhD degree in 1971. He worked on the design of circuits for television receivers for three years at Mullard (now Philips) Research Laboratories in Surrey.

He joined the Electrical and Electronic Engineering Department of Manchester Polytechnic in 1973 and has held the appointment of Principal Lecturer in Electronics since 1980. He is involved in research programmes and consultancies in the field of thick-film hybrid microelectronics. He is a corporate member of the Institution of Electrical Engineers and a member of ISHM (UK).

GEOFFREY W. GRIFFITHS is a graduate of London University in Physics and Mathematics and is currently the Technical Director of Newmarket Microsystems Ltd., a member of the Cambridge Electronics Industries group of companies. At Newmarket he has been involved in microelectronics since the mid-1950's and has been awarded two patents on semiconductor processing. Following this he initiated the company's hybrid activity in the mid-1960's which now forms the company's main product area. He was appointed a board member in 1970.

He has lectured extensively on thick film technology including in the People's Republic of China in 1973 and, under UNESCO sponsorship, in India in 1981. He also holds a Visiting Fellowship at the University of Hull and is a member of IEEE and ISHM (UK).

HAMISH T. LAW holds the degrees of BSc and PhD. He was formerly the Manager of the Hybrid Microelectronics Department of Ferranti Ltd., Edinburgh, where he built up a substantial activity in thin film technology. His current appointment is as the Director of the Bioengineering Unit at Princess Margaret Rose Orthopaedic Hospital.

Dr. Law is the immediate past chairman of ISHM (UK) and has represented ISHM (UK) on the European Liaison Committee of ISHM.

JIM T. LYNCH is a graduate of New College, Oxford gaining his BA in Chemistry and PhD in Surface Physics. He joined the Allen Clark Centre of the Plessey Co. Ltd. in 1970 and is now the deputy Q.A. Manager of the Quality Assurance division.

Dr. Lynch is concerned with reliability, failure analysis and product assessment of microelectronic devices particularly thick film hybrids. During the last eight years a principal duty has been acting for the Design Authority for Thick Film Hybrid Microcircuits, Plessey Telecommunications. This has involved vendor surveys of hybrid manufacturers, assessment of hybrids and the production of a Design Authority Manual dealing with all aspects of thick film hybrid design, production, procurement and test. He is also concerned with the development and maintenance of Quality Audit procedures within the Allen Clark Research Centre. He has done considerable work on literature survey and information retrieval by computer and worked on the committee commissioned by the U.K. Office of Scientific and Technical Information to produce the report on Information Research in the U.K.

PETER L. MORAN holds a BSc (Hons) in Electronics Science from the University of Southampton. He graduated under a U.K.A.E.A. student/graduate engineer training scholarship and at the U.K.A.E.A. was involved in studies on digital communication systems. In 1973 he joined the University of Technology, Loughborough as a lecturer in the Department of Electrical and Electronic Engineering and built up a substantial hybrid circuit activity. In 1980 he joined the then newly formed National

Microelectronics Research Centre at University College, Cork as Senior Scientist and as well as having interests in thick film methods, is also involved in more general aspects of interconnection technology.

He is a member of IEE and a member of ISHM (UK).

JOHN R. POLDEN holds a Bachelors degree in Electronic Engineering from the University of Southampton. He graduated under a U.K.A.E.A. student/graduate engineer scholarship and worked there as an electronic engineer. He then took a Masters Course at the London Business School.

Currently he is General Manager of the Microelectronics Division of Welwyn Electronics. The division specializes in hybrid circuitry and supplies the military, telecoms and automotive industries.

IAN D. SALISBURY entered industry with AEI working on semiconductor development and also industrial communication equipment. In 1960 he joined STC and was employed in developing various types of capacitors, in particular thin film devices. He was involved in the ITT hybrid circuit unit from its inception and became chief engineer for all film circuit production and development in 1977.

In 1981 he was appointed Sales and Marketing Manager for Engelhard specialising in capacitor and thick film products.

ANDREW G. SAUNDERS is head of the integrated circuit assembly and hybrid development group at British Telecom Research Laboratories, Martlesham Heath, Ipswich, U.K. After receiving his B.Sc. in physics from the University of East Anglia in 1969 and an M.Sc. from the University of Southampton in 1971 he worked for a period with GEC Semiconductors Ltd., before joining the (then) Post Office Research Department to work on the processing and assembly of high reliability transistors for submarine cable systems. In 1977 he started the thick film hybrid facility at BTRL, and now heads this unit as well as being responsible for the assembly of integrated circuits produced there.

Biography

F. NIHAL SINNADURAI is currently managing the research and development of Advanced Hardware Techniques at British Telecom Research Laboratories, dealing with interconnection technology and applications, and system reliability evaluation. He has specialised in semiconductor device technology and reliability, and has also contributed to the understanding of the behaviour and reliability of thick-film components and hybrid microcircuits. Dr. Sinnadurai has published and presented over 25 papers in these fields, and lectured on these topics internationally.

He is a Fellow of the Institute of Physics, a member of the International Society for Hybrid Microelectronics (ISHM), and has been active in advancing knowledge of hybrid microelectronics and interconnection techniques. He has organised and lectured at ISHM symposia and conferences in the U.K. and abroad.

Dr. Sinnadurai obtained his BSc Honours in Physics and MSc in the Physics of Semiconductor Devices at the University of London, and was awarded the PhD by the University of Southampton for his research and thesis on "Reliability Studies of Silicon Planar Devices and IMPATT Diodes".

JOHN C. TAYLOR received a BSc (Hons) in Electrical Engineering from Glasgow University in 1977 and underwent training at RSRE (MOD(PE)) Malvern during his degree course. After graduation he worked on radar signal processing and in 1978 he transferred to the Electronic Engineering Services Division within RSRE where he managed the Microelectronics Assembly Section of the hybrid unit. He now manages the hybrid unit which is involved in the development and prototype manufacture of a wide range of microcircuits within RSRE.

He is a member of ISHM (UK) and an Associate Member of the IEE.

BRIAN WALTON is a physicist who has been actively involved in research and development related to electronic components and microelectronics for over twenty years. After holding a number of posts in industry he joined ERA Technology Ltd., the Leatherhead based independent research and development organisation for the electrical industry, where he is currently

Manager of the Electronics Technology Department. He has been responsible for the development of a number of novel thick film materials and processes and for the objective assessment of commercially available systems. Current centre of interest is new packaging and interconnection methods and materials for dense circuitry.

BRIAN C. WATERFIELD has an HNC in Mechanical and Electrical Engineering from Rugby College of Technology and subsequently trained as a ceramics engineer.

He was one of the original designers of the Integral Substrate Packaging concept and produced all ceramic packages using thick film methods in 1973.

Currently he is involved in numerous packaging projects as a consultant, mostly using glass, glass/metal and ceramic technologies.

He is the proprietor of Microelectronic and Industrial Services, a supplier of packages and related materials.

Index

A-D Converter, 48-152
AQL, 200
Accelerated ageing, 163, 194
Acceleration Test (see centrifuge test)
Active Filters, 90, 147
Active trimming, 33
Add-on components, 33, 55, 126
Adhesives, 58, 59, 99, 171
Alumina, 26, 50, 54
Arrhenius, 164
Automotive hybrids, 140

BS 9450, 23, 191
Barium Titanate, 31
Bismuth Ruthenate, 47

CAD, 24, 35, 135
CQC, 191
CTR, 197, 199
Capability approval (see BS 9450)
Capacitor ageing, 101
Capacitor discharge welding, 111
Capacitors, 55
Cardiac pacemakers, 18, 146
Cassette loading, 125
Centrifuge test, 196

Ceramic packages, 116
Chip carriers, 33, 60, 90, 101, 154, 163, 178
Conductor end effect, 31
Connectors, 13, 17

D 4500, 167, 201
Damp heat cycling, 195
Devitrification, 49
Drying (thick film pastes), 31

Electromigration, 164
Electroplating, 70

Fibre optic hybrids, 150, 160
Furnace, 33

Gador, 73
Gaelic, 70
Glass-metal seal, 113
Glass sealing, 111

Hearing Aids, 147
Hermeticity, 120, 195

IBM, 14
Integral substrate package, 87,

116, 171
Integrated circuits, 5
  attachment, 58, 61, 127
  i/o, 11, 12
  packaging, 8, 56, 108
  reliability, 8, 164
  semi-custom, 11, 154
  testing, 8, 161

Kovar, 115

Laser scribed substrates, 125, 132
Lead frame, 60, 133, 140
Leak detection, 195

Medical hybrids, 146

Nitrogen firing, 51

Philips, 17, 178
Piezo resistive effect, 144
Plastic coated hybrids 87, 109, 174, 185
Platform package, 115, 133
Polymer substrates, 52, 55, 132
Polymer thick film, 52
Printed circuitry, 12, 157
Projection welding, 111
Protective finishes, 118

Quality assurance, 184

Reactive Bonding, 29
Reflow soldering, 58, 129
Resist coating, 74
Rework, 37, 105
Rubylith, 29
Ruthenium Oxide, 29, 144

SO Package, 33, 60, 90, 137, 161
Screen printer, 31

Seam welding, 113
Sheet resistivity, 28
Shock test, 198
Solid sidewall package, 115, 133
Sputtering, 65

TAB, 108
TCR, 29, 44
Telecommunications hybrids, 150, 160
Telemetry hybrids, 147
Testing, 38, 103, 133, 161
Thermal cycling, 185, 195, 199
Thermal management, 38, 99
Thermal mismatch, 12, 103, 184
Thermal printheads, 145
Thermistors, 144
Thick film, 14, 22
  conductors, 14, 28, 43, 45, 90, 185
  crossover, 94
  inductors, 151
  insulators, 14, 31, 47, 137
  integration, 23, 160
  layout, 35
  pastes, 28, 41
  reliability, 23, 97, 146, 159, 163, 166, 170
  resistor trimming, 14, 33, 96, 126
  resistors, 14, 29, 44, 96, 137
  screens, 19, 26
Thin film, 13, 55, 64
  etching, 75
  ionic degradation, 69
  microwave, 20, 151
  packaging, 13, 87
  preparation, 65
  reliability, 13, 80
  resistor trimming, 77
  resistor calculation, 71

resistors, 13
uncommitted substrate, 67
Transduces (hybrid), 144

Uncommitted thin film substrates, 67

Vibration tests, 185, 197
Vitreous enamelled steel, 51, 55

Weld scaling, 111
Wire bonding, 59, 61, 94, 128, 189

Yield, 6, 35, 161